T0276474

Encyclopedia of Pesticides: Management and Sustainable Development

Volume II

Encyclopedia of Pesticides: Management and Sustainable Development
Volume II

Edited by **Edwin Tan**

New York

Published by Callisto Reference,
106 Park Avenue, Suite 200,
New York, NY 10016, USA
www.callistoreference.com

Encyclopedia of Pesticides: Management and Sustainable Development
Volume II
Edited by Edwin Tan

International Standard Book Number: 978-1-63239-278-7 (Hardback)

Printed in the United States of America.

Contents

Preface

This book covers matters regarding pesticides as well as biopesticides with a special focus on the associated subjects of pesticides management and sustainable development. An analysis on the ecological effects of pesticides, on the levels of contamination, biodegradation of pesticides, pesticides-soil interactions, pesticides management options, and on few methods of pesticides application, reporting data accumulated from across the globe has been presented in this book. The topics covered in the book will be beneficial to students, researchers, government officials and members of the civil society working in the field.

Significant researches are present in this book. Intensive efforts have been employed by authors to make this book an outstanding discourse. This book contains the enlightening chapters which have been written on the basis of significant researches done by the experts.

Finally, I would also like to thank all the members involved in this book for being a team and meeting all the deadlines for the submission of their respective works. I would also like to thank my friends and family for being supportive in my efforts.

Editor

Pesticides Management and Sustainable Development

Ecological Effects of Pesticides

Deepa T.V.[1], G. Lakshmi[1], Lakshmi P.S.[1] and Sreekanth S.K.[2]

[1]Amrita School of Pharmacy, Amrita Viswa Vidyapeetham
[2]St.James Medical Acadamy,Calicut University
India

1. Introduction

Pesticides which are used for preventing or destroying pest is having more negative impact on our ecological system when compared to its desired action. Pesticides are carried by wind to other areas and make them contaminate. Pesticides are also causing water pollution and some pesticides are persistent organic pollutants which contribute to soil contamination. (Rockets & Rusty,2007).

The amount of pesticide that migrates from the intended application area is influenced by the particular chemical's properties: its propensity for binding to soil, its vapor pressure, its water solubility, and its resistance to being broken down over time (Tashkent, 1998) Some pesticides contribute to global warming and the depletion of the ozone layer.

2. Water

Pesticides were found to pollute every source of water including wells.(Gilliom et al.,2007).Pesticide residues have also been found in rain and groundwater.(Kellogg et al.,2000).Pesticide impacts on aquatic systems are often studied using a hydrology transport model to study movement and fate of chemicals in rivers and streams Studies by the UK government showed that pesticide concentrations exceeded those allowable for drinking water in some samples of river water and groundwater. (Bingham,2007).

The main routes through which pesticides reach the water are:

1. It may drift outside of the intended area when it is sprayed.
2. It may percolate, or leach, through the soil.
3. It may be carried to the water as runoff.
4. It may be spilled accidentally or through neglect. (States of Jersey, 2007).

They may also be carried to water by eroding soil. (Papendick et al.,1986) Factors that affect a pesticide's ability to contaminate water include its water solubility, the distance from an application site to a body of water, weather, soil type, presence of a growing crop, and the method used to apply the chemical. Maximum limits of allowable concentrations for individual pesticides in public bodies of water are set by the Environmental Protection Agency in the US (Pedersen,1997).

3. Soil

Many of the chemicals used in pesticides are persistent soil contaminants, whose impact may endure for decades and adversely affect soil conservation (U.S. Environmental

Protection Agency,2007).The use of pesticides decreases the general biodiversity in the soil. Not using the chemicals results in higher soil quality verified needed, (Johnston,1986) with the additional effect that more organic matter in the soil allows for higher water retention.(Kellogg et al.,2000).This helps increase yields for farms in drought years, when organic farms have had yields 20-40% higher than their conventional counterparts.(Lotter et al.,2003) A smaller content of organic matter in the soil increases the amount of pesticide that will leave the area of application, because organic matter binds to and helps break down pesticides.(Kellogg et al.,2000).

4. Air

Pesticides can contribute to air pollution . Pesticide drift occurs when pesticides suspended in the air as particles are carried by wind to other areas, potentially contaminating them. (Kellogg et al.,2000). Volatile pesticides applied to crops will volatilize and are blown by winds to nearby areas posing a threat to wildlife.(Reynolds,1997). Sprayed pesticides or particles from pesticides applied as dusts may travel on the wind to other areas, or pesticides may adhere to particles that blow in the wind, such as dust particles.(National Park Service,2006). Compared to aerial spraying ground spraying produces less pesticide drift. (U.S. Environmental Protection Agency,PR,2007) Farmers can employ a buffer zone around their crop, consisting of empty land or non-crop plants such as evergreen trees to serve as windbreaks and absorb the pesticides, preventing drift into other areas.

5. Effects on biota

5.1 Plants
Nitrogen fixation, which is required for the growth of higher plants, is hindered by pesticides in soil. The insecticides DDT, methyl parathion, and especially pentachlorophenol have been shown to interfere with legume-rhizobium chemical signaling. Reduction of this symbiotic chemical signaling results in reduced nitrogen fixation and thus reduces crop yields(Rockets&Rusty,2007). Root nodule formation in these plants saves the world economy $10 billion in synthetic nitrogen fertilizer every year.(Fox et al.,2007).
Pesticides can kill bees and are strongly implicated in pollinator decline, the loss of species that pollinate plants, including through the mechanism of Colony Collapse Disorder, (Wells, 2007) in which worker bees from a beehive or Western honey bee colony abruptly disappear. Application of pesticides to crops that are in bloom can kill honeybees,(Cornell University,2007) which act as pollinators. The USDA and USFWS estimate that US farmers lose at least $200 million a year from reduced crop pollination because pesticides applied to fields eliminate about a fifth of honeybee colonies in the US and harm an additional 15%..(Rockets & Rusty,2007),

5.2 Animals
Pesticides inflict extremely widespread damage to biota, and many countries have acted to discourage pesticide usage through their Biodiversity Action Plans.Animals may be poisoned by pesticide residues that remain on food after spraying, for example when wild animals enter sprayed fields or nearby areas shortly after spraying. (Palmer et al.,2007).

Fumigators walking down a street in the Sultan Mosque area of Singapore and spraying a pesticide to rid the area of mosquitoes

Widespread application of pesticides can eliminate food sources that certain types of animals need, causing the animals to relocate, change their diet, or starve. Poisoning from pesticides can travel up the food chain; for example, birds can be harmed when they eat insects and worms that have consumed pesticides. Some pesticides can cause bioaccumulation, or build up to toxic levels in the bodies of organisms that consume them over time, a phenomenon that impacts species high on the food chain especially hard. (Cornell University,2007).

Pesticides can affect animal reproduction directly, as evidenced by the deleterious effect of the persistent organochlorine insecticides on reproduction in raptors and other birds. Eggshell thinning due to the uptake of organochlorine insecticides that affect calcium (Ca) metabolism has been observed in predacious birds (Keith *et al.*, 1970; Newton *et al.*, 1986; Wiemeyer *et al.*, 1986; Opdam *et al.*, 1987). Fish-eating birds are more severely affected than terrestrial predatory birds, because the fish-eating birds acquire more pesticides via their food chain than the other predators (Pimentel, 1971; Littrell, 1986).

Pesticides can also affect reproduction in the invertebrates; for example, sublethal doses of DDT, dieldrin, and parathion increased egg production by the Colorado potato beetle by 50, 33 and 65 per cent, respectively, after two weeks (Abdallah, 1968). The herbicide 2,4,5-T was found to reduce the reproduction of soil-inhabiting Collembola (Eijsackers, 1978). Populations of invertebrates with high rates of increase can recover stable populations much more rapidly than those of bird and mammal populations (Pimentel and Edwards, 1982).

5.2.1 Human

Pesticides can enter the human body through inhalation of aerosols, dust and vapor that contain pesticides; through oral exposure by consuming food and water; and through dermal exposure by direct contact of pesticides with skin.(Department of Pesticide Regulation,2008).Pesticides are sprayed onto food, especially fruits and vegetables, they secrete into soils and groundwater which can end up in drinking water, and pesticide spray can drift and pollute the air.

There is increasing anxiety about the importance of small residues of pesticides, often suspected of being carcinogens or disrupting endocrine activities, in drinking water and food. In spite of stringent regulations by international and national regulatory agencies, reports of pesticide residues in human foods, both imported and home-produced, are numerous.

Over the last fifty years many human illnesses and deaths have occurred as a result of exposure to pesticides, with up to 20,000 deaths reported annually. Some of these are suicides, but most involve some form of accidental exposure to pesticides, particularly among farmers and spray operators in developing countries, who are careless in handling pesticides or wear insufficient protective clothing and equipment. Moreover, there have been major accidents involving pesticides that have led to the death or illness of many thousands. One instance occurred in Bhopal, India, where more than 5,000 deaths resulted from exposure to accidental emissions of methyl isocyanate from a pesticide factory.

The effects of pesticides on human health are more harmful based on the toxicity of the chemical and the length and magnitude of exposure.(Lorenz & Eric,2009). Farm workers and their families experience the greatest exposure to agricultural pesticides through direct contact with the chemicals. But every human contains a percentage of pesticides found in fat samples in their body. Children are most susceptible and sensitive to pesticides due to their small size and underdevelopment. (Department of Pesticide Regulation,2008). The chemicals can bioaccumulation in the body over time.

Exposure to pesticides can range from mild skin irritation to birth defects, tumors, genetic changes, blood and nerve disorders, endocrine disruption, and even coma or death. (Lorenz & Eric,2009)

5.2.2 Aquatic life
A major environmental impact has been the widespread mortality of fish and marine invertebrates due to the contamination of aquatic systems by pesticides. This has resulted from the agricultural contamination of waterways through fallout, drainage, or runoff erosion, and from the discharge of industrial effluents containing pesticides into waterways. Historically, most of the fish in Europe's Rhine River were killed by the discharge of pesticides, and at one time fish populations in the Great Lakes became very low due to pesticide contamination

Fish and other aquatic biota may be harmed by pesticide-contaminated water. Pesticide surface runoff into rivers and streams can be highly lethal to aquatic life, sometimes killing all the fish in a particular stream.(Toughill,1999).

Application of herbicides to bodies of water can cause fish kills when the dead plants rot and use up the water's oxygen, suffocating the fish. Some herbicides, such as copper sulfite, that are applied to water to kill plants are toxic to fish and other water animals at concentrations similar to those used to kill the plants, Repeated exposure to sub lethal doses of some pesticides can cause physiological and behavioral changes in fish that reduce populations, such as abandonment of nests and broods, decreased immunity to disease, and increased failure to avoid predators,(Helfrich et al.,1996).

Application of herbicides to bodies of water can kill off plants on which fish depend for their habitat.(Helfrich et al.,1996).Pesticides can accumulate in bodies of water to levels that kill off zooplankton, the main source of food for young fish.(Pesticide Action Network North America,1999).Pesticides can kill off the insects on which some fish feed, causing the fish to travel farther in search of food and exposing them to greater risk from predators.The faster a given pesticide breaks down in the environment, the less threat it poses to aquatic life. Insecticides are more toxic to aquatic life than herbicides and fungicides. (Helfrich et al.,1996).

5.2.3 Birds
Pesticides had created striking effects on birds, those in the higher trophic levels of food chains, such as bald eagles, hawks, and owls. These birds are often rare, endangered, and susceptible to pesticide residues such as those occurring from the bioconcentration of organochlorine insecticides through terrestrial food chains. Pesticides will also kill grain- and plant-feeding birds, and the elimination of many rare species of ducks and geese has been reported. Populations of insect-eating birds such as partridges, grouse, and pheasants have decreased due to the loss of their insect food in agricultural fields through the use of insecticides.

Bees are extremely important in the pollination of crops and wild plants, and although pesticides are screened for toxicity to bees, and the use of pesticides toxic to bees is permitted only under stringent conditions, many bees are killed by pesticides, resulting in the considerably reduced yield of crops dependent on bee pollination.

Bald eagles are common examples of nontarget organisms that are impacted by pesticide use. Rachel Carson's landmark book Silent Spring dealt with the of loss of bird species due to bioaccumulation of pesticides in their tissues. There is evidence that birds are continuing to be harmed by pesticide use. In the farmland of Britain, populations of ten different species of birds have declined by 10 million breeding individuals between 1979 and 1999, a phenomenon thought to have resulted from loss of plant and invertebrate species on which the birds feed. Throughout Europe, 116 species of birds are now threatened. Reductions in bird populations have been found to be associated with times and areas in which pesticides are used. (Kerbs et al.,1999) In another example, some types of fungicides used in peanut farming are only slightly toxic to birds and mammals, but may kill off earthworms, which can in turn reduce populations of the birds and mammals that feed on them. (Palmer et al.,2007).

Some pesticides come in granular form, and birds and other wildlife may eat the granules, mistaking them for grains of food. A few granules of a pesticide is enough to kill a small bird. (Palmer et al.,2007).

The herbicide parquet, when sprayed onto bird eggs, causes growth abnormalities in embryos and reduces the number of chicks that hatch successfully, but most herbicides do not directly cause much harm to birds. Herbicides may endanger bird populations by reducing their habitat.(U.S. Environmental Protection Agency,2007).

A bird that died as a result of pesticide use. (U.S. EPA. Reproduced by permission.)

6. Threatening reports on hazardous effects of pesticides

Endosulfan is a harmful insecticide , it causes several health hazards in human beings.endosulfan was aerial sprayed on cashew plantations in india especially in northern parts of kerala for more than 20 years.the terrain was unsuitable for aerial spraying considering the relatively high rainfall and its geological structure. unsual diseases and even deaths were observed in and around the region.

Endosulfan is a chlorinated hydrocarbon insecticide of the cyclodiene subgroup which act as a contact poison in wide variety of insects and mice is primarily used on food crops like tea,fruits,vegetable and grains. exposure to endosulfan will result from ingestion of contaminated food.it does not easily dissolve in water and tranport is likely occur if it attached to soil particles in surface runoff.endosulfan residues have been found in numerous food products at very low concentration.

Endosulfan is rapidly degraded and eliminated in mammals with very little absorption in gastrointestinal tract. in these areas, where aerial spraying was done lot of children who have exposed are considered to be living martyr.

Studies consistently show thatendosulfanis highly poisonous and easily causes death and severe acute and chronic toxicity to various organ system including mental impairment , neurologic disturbances , immunotoxicity and reproductive toxicity and most of the new born were physically handicapped and showing epilepsy.

Classified by the us environmental protection agency as highly hazardous endosulfan was at the centre of controversy in the philippinese in 1990's.(Nishand .P,2006)

7. Alternative methods foreliminating pesticides

7.1 Diversified planting

A common practice among home gardeners is to plant a single crop in a straight row. This encourages pest infestation because it facilitates easy travel of an insect or disease from one host plant to another. By intermingling different types of plants and by not planting in straight rows, an insect is forced to search for a new host plant thus exposing itself to predators. Also, this approach corresponds well with companion planting. .(Ann R. Waters ,2011)

7.2 Low toxicity pesticides

Formulated, biodegradable pest-control substances are commercially available. Although these products are pesticides, they have low toxicity to mammals and do not last long in the environment. The local County Extension Service can provide information on these and other pesticide products.(Ann R. Waters ,2011)

Many alternatives are available to reduce the effects pesticides have on the environment. There are a variety of alternative pesticides such as manually removing weeds and pests from plants, applying heat, covering weeds with plastic, and placing traps and lures to catch or move pests. Pests can be prevented by removing pest breeding sites, maintaining healthy soils which breed healthy plants that are resistant to pests, planting native species that are naturally more resistant to native pests, and use biocontrol agents such as birds and other pest eating organisms. (National Audubon Society,2003).

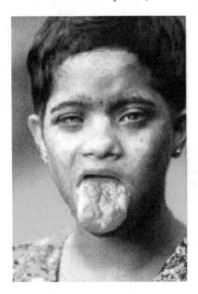

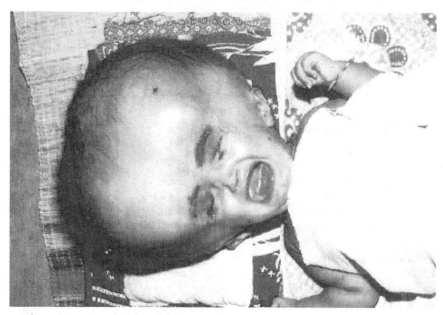

Photos of some victims of endosulphan tragedy

8. References

Abdallah, M. D. (1968) The effect of sublethal dosages of DDT, parathion, and dieldrin on oviposition of the Colorado potato beetle (*Leptinotarsa decemlineata* Say) (*Coleoptera: Chrysomelidae*). Bull. Entomol. Soc. Egypt Econ. Ser. 2, 211-217.

Ann R. Waters,Carmen Valentin ,Alternatives to pesticides Outreach Coordinator Pesticide Control Program,2011.

Bingham, S (2007), Pesticides in rivers and groundwater. Environment Agency, UK.

States of Jersey (2007), Environmental protection and pesticide use.

Cornell University. Pesticides in the environment. Pesticide fact sheets and tutorial, module 6. Pesticide Safety Education Program,2007.

Department of Pesticide Regulation (2008), "What are the Potential Health Effects of Pesticides?" Community Guide to Recognizing and Reporting Pesticide Problems. Sacramento, CA. Pages 27-29.

Eijsackers, H. (1978) Side effects of the herbicide 2,4,5-T on reproduction, food consumption, and moulting of the springtail *Onychiurus quadriocellatus* Gisin (Collembola). Z. Angew. Entomol. 85, 341-360.

Fox, JE, Gulledge, J, Engelhaupt, E, Burrow, ME, and McLachlan, JA (2007). "Pesticides reduce symbiotic efficiency of nitrogen-fixing rhizobia and host plants". *Proceedings of the National Academy of Sciences of the USA*

Gilliom, RJ, Barbash, JE, Crawford, GG, Hamilton, PA, Martin, JD, Nakagaki, N, Nowell, LH, Scott, JC, Stackelberg, PE, Thelin, GP, and Wolock, DM (February 15, 2007),

The Quality of our nation's waters: Pesticides in the nation's streams and ground water, 1992–2001. Chapter 1, Page 4. US Geological Survey.

Helfrich, LA, Weigmann, DL, Hipkins, P, and Stinson, ER (June 1996), Pesticides and aquatic animals: A guide to reducing impacts on aquatic systems. Virginia Cooperative Extension.

Johnston, AE (1986). "Soil organic-matter, effects on soils and crops". Soil Use Management 2: 97–105.

Keith, J. O., Woods, L. A. and Hunt, E. C. (1970) Reproductive failure in brown pelicans on the Pacific coast. *Transactions of the 35th North American Wildlife and Natural Resources Conference,* Wildlife Management Institute, Washington, DC, pp. 56-63.

Kellogg RL, Nehring R, Grube A, Goss DW, and Plotkin S (February 2000), Environmental indicators of pesticide leaching and runoff from farm fields. United States Department of Agriculture Natural Resources Conservation Service.

Kerbs JR, Wilson JD, Bradbury RB, and Siriwardena GM (August 12, 1999), The second silent spring. Commentary in *Nature,* Volume 400, Pages 611-612.

Littrell, E. E. (1986) Shell thickness and organochlorine pesticides in osprey eggs from Eagle Lake, California. *Calif. Fish Game* 72, 182-185.

Lotter DW, Seidel R, and Liebhardt W (2003). "The performance of organic and conventional cropping systems in an extreme climate year". American Journal of Alternative Agriculture 18: 146–154.

Lorenz, Eric S. "Potential Health Effects of Pesticides." Ag Communications and Marketing (2009). Pages 1-8.

Newton, I., Bogan, J. A. and Rothery, P. (1986) Trends and effects of organochlorine compounds in sparrowhawk eggs. *J. Appl. Ecol.* 23, 461-478.

National Park Service. US Department of the Interior. (August 1, 2006), Sequoia & Kings Canyon National Park: Air quality -- Airborne synthetic chemicals.

Nishand .P(2006),Effects of endosulphan on human beings

Odum, E. P. (1971) *Fundamentals of Ecology* (3rd edn), Saunders, Philadelphia. Opdam, P., Burgers, J. and Müskens, G. (1987) Population trend, reproduction, and pesticides in Dutch sparrowhawks following the ban on DDT. *Ardea* 75(2), 205-212.

Papendick RI, Elliott LF, and Dahlgren RB (1986), Environmental consequences of modern production agriculture: How can alternative agriculture address these issues and concerns? *American Journal of Alternative Agriculture,* Volume 1, Issue 1, Pages 3-10.

Palmer, WE, Bromley, PT, and Brandenburg, RL. Wildlife & pesticides - Peanuts. North Carolina Cooperative Extension Service. Retrieved on 2007-10-11.

Pedersen, TL (June 1997), Pesticide residues in drinking water

Pesticide Action Network North America (June 4, 1999), Pesticides threaten birds and fish in California.

Pimentel, D. (1971) *Ecological Effects of Pesticides on Non-Target Species,* Executive Office of the President, Office of Science and Technology, Washington, DC, 220pp.

Pimentel, D. and Edwards, C. A. (1982) Pesticides and ecosystems. *BioScience* 32, 595-600.

Reynolds, JD (1997), International pesticide trade: Is there any hope for the effective regulation of controlled substances? *Florida State University Journal of Land Use & Environmental Law*, Volume 131.

Rockets, Rusty (June 8, 2007), Down On The Farm? Yields, Nutrients And Soil Quality.

Take Action! How to Eliminate Pesticide Use." (2003) National Audubon Society. Pages 1-3.

Tashkent (1998), Part 1. Conditions and provisions for developing a national strategy for biodiversity conservation. Biodiversity Conservation National Strategy and Action Plan of Republic of Uzbekistan. Prepared by the National Biodiversity Strategy Project Steering Committee with the Financial Assistance of The Global Environmental Facility (GEF) and Technical Assistance of United Nations Development Programme (UNDP).

Toughill K (1999), The summer the rivers died: Toxic runoff from potato farms is poisoning P.E.I. Originally published in *Toronto Star Atlantic Canada Bureau*.

U.S. Environmental Protection Agency (2007), Sources of common contaminants and their health effects.

US Environmental Protection Agency (September 11th, 2007), Pesticide registration (PR) notice 2001-X Draft: Spray and dust drift label statements for pesticide products.

Wells, M (March 11, 2007). "Vanishing bees threaten US crops". www.bbc.co.uk (*BBC News*).

Wiemeyer, S. N., Porter, R. D., Hensler, G. L. and Maestrelli, J. R. (1986) *DDE, DDT & Dieldrin: Residues in American Kestrels and Relations to Reproduction*, Fish and Wildlife Technical Report 6, US Department of the Interior, Fish and Wildlife Service.

2

Increasing IPM Knowledge Through FFS in Benin

Trine Lund and Hafizur Rahman
Norwegian University of Life Sciences
Norway

1. Introduction

Over the next two generations 4 billion more people will live in cities, increasing the proportion of the urban population from 50 to 80 per cent of the total world population (NRC, 1999). Thus a sustainable development needs to focus on meeting the needs of an increasing human population, reducing poverty and hunger while at the same time sustaining the life support systems of the planet (NRC, 1999). While the Green Revolution technologies enabled extensive monocultures and higher yields through improved seeds, chemical fertilizers and synthetic pesticides, biodiversity in and around the agro-ecosystems have been reduced, causing the loss of natural pest and disease control (Gallagher et al., 2005). This has increased the need for synthetic pesticides in the agricultural sector to the current global use of 2.56 billion kgyr^{-1} (Pretty, 2008), with associated negative effects for humans and the ecosystem becoming evident. While the externalities of pesticides in rice systems in China cost $1.4 billion per year through adverse effects on biodiversity and people's health (Norse et al., 2001), the annual mortality rate due to pesticides in the remote Ecuadorian highlands is among the world's highest, 21 per 100 000 people (Sherwood et al., 2005). On the other hand, in the Philippines, agricultural systems that do not use any synthetic pesticides experience higher net social benefits due to reduced illnesses among farmers and their families, and associated lower medical costs (Rola & Pingali, 1993, Pingali & Roger, 1995). According to FAO and ILO estimates, 2 to 5 million agricultural workers yearly experience severe pesticide poisoning and related illnesses of which 40 000 are lethal (FAO & ILO, 2009). However, pesticide poisoning incidents are often underreported, as indicated by a study among farmers in Senegal, Mali and Benin, where over 80% of the respondents faced adverse effects after spraying pesticides, to the extent of blurred vision, unconsciousness or severe dizziness, but only 2% sought medical attention for these symptoms (Thiam & Touni, 2009). Thus recent studies, where 4% to 9% of the surveyed farmers reported poisoning incidents the last year, estimate a yearly 25 – 45 million poisoning cases (Kishi, 2005).

Africa only accounts for 4% of pesticides used globally, an estimated 75-100 metric tons of pesticide active ingredient (compared to 350,000 tons in Europe), and average pesticide use per hectare cultivated land in Africa (1.23kg/ha) is very low compared to Latin America and Asia (7.17kg/ha and 3.12kg/ha respectively) (Thiam & Touni, 2009). Still, the risks and impacts associated with synthetic pesticides are not necessarily lower in Africa as many of the pesticides used in the continent are adulterated, poor quality and unlabelled and application and handling practices are often highly unsafe (Thiam & Touni, 2009, Lund et al., 2010). In a study among farm families in Senegal and Benin, the number of pesticide poisoning incidents were 619 and 84 respectively of which 16% and 23% were fatal (Thiam & Touni, 2009). Only 2% of the farmers in the studied villages used a full set of protective gear (gloves, boots, and masks or glasses), showing how unavailability and impracticality of protective gear has an enormous impact on poisoning levels and farmers' health (ibid.). The farmers' families and communities also experienced negative effects of pesticides, like accidental poisonings, because pesticides are often stored freely available in kitchen or bedrooms, empty pesticide containers are reused for food and drinks and pesticides are purchased in non-original containers (ibid.). Governments tend to focus on the ones handling pesticides directly assuming that those face the highest poisoning risk, but data from Benin, Senegal, Ethiopia (ibid.), Ecuador (Cole et al., 2002) and India (Mancini et al., 2005) shows high frequencies of pesticide poisonings among women and children even though they are generally not applying pesticides.

The World Health Organization (WHO) has classified pesticides in Ia (extremely hazardous), Ib (highly hazardous), II (moderately hazardous), III (slightly hazardous) and unlikely hazardous.There is increasing pressure from and work done by government regulators and civil society to prohibit the use of the Class Ia and Ib pesticides (Thiam & Touni, 2009). The frequent incidents of acute and fatal poisonings from Class II pesticides in Benin and Senegal, illustrates the dangerous effects of even "moderately hazardous" pesticides in conditions of poverty and poor education, showing that also Class II and Class III compounds (e.g. malathion) should be considered for restrictions (ibid.). The Persistent Organic Pollutant Endosulfan (Class II), which has been widely used in West African cotton growing areas, was banned by governments in the region in 2008, as it had been associated with acute and fatal poisonings (ibid.). Pesticides cause long-term health problems such as birth defects and cancers (Lichtenberg, 1992) and several studies link pesticide exposure to respiratory problems, memory disorders (Arcury et al., 2003), dermatologic conditions (O'Malley, 1997), anxiety, depression (Beseler et al., 2008), and neurological disorders (Ascherio et al., 2006). WHO estimated that long-term exposure to pesticides may result in a yearly 735,000 people globally suffering specific chronic defects and 37,000 cases of cancer (WHO, 1990). Thus also health Ministries in six Central American countries have proposed a regional ban on the Class II pesticides endosulfan, paraquat and chlorpyrifos, in addition to pesticides in class 1a and 1b, based on results from an eight year poisoning surveillance program (Rosenthal, 2005).

Pesticide residues may interfere with the legume-rhizobium chemical signalling reducing nitrogen fixation and crop yields (Fox et al., 2007), and over 95% of applied herbicides and 98% of insecticides reach other destinations than their target, including non-target species, water, air, bottom sediments, and food (Miller & Spoolman, 2009). The use of synthetic pesticides among vegetable producers in urban and peri-urban areas of West-Africa has increased to the extent that certain insect pests have developed resistance to the pesticides (Atcha-Ahowé et al., 2009). Additional negative effects are increasing insecticide resistance in insect vectors due to the leakage of insecticides to mosquito breeding sites (Akogbeto et al., 2008), to the extent that insecticide resistance in Anopheles mosquitoes is threatening the success of malaria control programs (Djouaka et al., 2005), and pesticide resistance in target pests has made pest resurgence a common phenomenon in cotton, vegetables, rice and fruit crops production systems (Lim, 1992). Recent research also indicates that toxic compounds may be dispersed to even remote areas via atmospheric deposition (Rosendahl et al., 2009). In West-Africa, agricultural land is being degraded by poor agricultural practices and the use of chemical products (Pimbert et al., 2010). Still, African farmers often use credit to buy inputs such as seeds, fertilizers and synthetic pesticides at high costs, making them dependent on good yields to break even and manage their debts (Williamson, 2003). In this situation, many may get caught in the pesticide treadmill where they do not dare to reduce the use of synthetic pesticides for fear of yield loss. While the externally funded West-African agricultural research system increasingly focus on the use of imported fertilizers and pesticides, the use of traditional seeds and organic manure is declining and small-scale producers have felt lack of citizen control over knowledge production (Pimbert et al., 2010). January 2006, the local government of Sikasso in Mali hosted the Citizen Space for Democratic Deliberation on GMOs and the Future of Farming in Mali where local farmers made policy recommendations based on expert evidence from various sources (ibid.). The farmers requested a re-orientation of public research from the current focus on input-intensive farming and GM seeds, towards ecological farming not requiring chemical inputs, improved local seeds and landraces, regeneration of local markets and food systems, supporting small-scale producers. They also suggested that farmers set the research objectives and called for more exchange between farmers and researchers as well as the development of new Integrated Pest Management (IPM) strategies and training in these strategies taking local knowledge into account (ibid.). Also the recent International Assessment of Agricultural Knowledge, Science and Technology for Development (IAASTD) panel supported by over 400 experts under the co-sponsorship of the FAO, GEF, UNDP, UNEP, UNESCO, the World Bank and WHO, called for new farmer-scientist partnerships, to improve understanding of agro-ecology, i.e. by IPM, and develop an integrated agro-ecosystem and human health approach to enhance food security and safety, stating that: "The way the world grows its food will have to change radically to better serve the poor and hungry if the world is to cope with growing population and climate change while avoiding social breakdown and environmental collapse" (McIntyre et al., 2009). A recent UNEP and UNCTAD report based on 24 African countries states a 100% yield increase with organic or near-organic practices, concluding that organic practices in Africa outperformed industrial, chemical-intensive conventional farming and in addition improved soil fertility, water retention and draught resistance, making it a promising approach for food security in the continent (UNEP & UNCTAD, 2008).

As the 'top-down' recommendations for pest control, have often failed to reduce pest damage, pesticide use or enable farmers to learn about IPM (Williamson, 1998), there is a need for new ways of learning (Orr, 1992, Bentley et al., 2003, Liebelin et al., 2004, Bawden, 2005, Chambers, 2005). One learning method focusing on the farmers' own development is the farmer field school (FFS), which is increasingly being used to promote IPM. Since IPM FFS was introduced by the Global IPM Facility in West Africa (Ghana) in 1996, it has spread to over a dozen countries, from Senegal to South Africa (WB, 2004). IPM-FFS has been adopted in the national policies in Mali, Burkina Faso and Senegal, and IPM curricula, initially made for rice, have been developed for other crops including vegetables (ibid.). Community IPM, which has been used to increase the community involvement and adaptation of IPM in Asia for 15 years, is now being tested in Africa, including Burkina Faso, Ghana, Mali and Senegal in West Africa (ibid.). In 2003, the International Institute of

Tropical Agriculture (IITA) initiated the project 'Healthy Vegetables through Participatory Integrated Pest Management (IPM) in Peri-urban Gardens of Benin' (hereafter referred to as the project) to enhance farmers' efforts to produce quality vegetables through informed decisions on the choice and use of IPM options (James et al., 2006). The project was unique as it was the first time that a vegetable FFS was conducted by IITA in West Africa. This chapter will discuss the effects of IPM-FFS in pest management, including the IITA vegetable IPM-FFS as a case study.

2. The use of IPM and FFS in pest management

2.1 IPM and FFS

Mature ecosystems' state of dynamic equilibrium buffers against large shocks and stress, but modern agro-ecosystems have weak resilience (ability to resist stress and shocks) (Holling et al., 1998, Folke, 2006, Shennan, 2008). Thus developing sustainability requires a focus on structures and functions to improve the resilience, such as increasing the biodiversity to recreate natural pest and disease control, rather than seeking to eliminate those populations (ibid). In ecosystems, multi-trophic interactions are vital (Shennan, 2008). For example foliar herbivory in grasslands impact the functions of soil food webs (Wardle, 2006), which, together with changed nutrient dynamics, in turn affect the plant attractiveness to herbivores (Awmack & Leather, 2002, Beanland et al., 2003). Also due to the crops' systemic defense mechanisms above-ground attack may trigger responses to below-ground attack and vice versa (Bruce & Pickett, 2007). These complex crop–weed–disease–pest interactions imply that farm practices such as crop rotation, tillage, pesticides and fertilizers affect disease incidence, weed and pest populations (Bruce & Pickett, 2007), while practices like utilizing nitrogen fixing legumes, natural enemies for pest management and applying zero-tillage may enhance the sustainability (Pretty, 2008). As the importance of the complex interrelationships between the crop, weed, disease and pests is increasingly documented, the reductionist view of applying a synthetic pesticide to fix a specific pest problem is being questioned (NRC, 2010). Thus the pressure for pesticide reductions has influenced research to shift its focus towards non-chemical alternatives (ibid.). Increased emphasize is being paid to the approach Integrated Pest Management (IPM), which appreciates the complexity of the agro-ecosystem and "utilizes all suitable techniques in the socioeconomic, environment and population dynamics context of farming systems in a compatible manner to maintain pest population levels below those causing economic injury" (Dent, 1995: 1). IPM has proved able to reduce or eliminate the use of synthetic pesticides while improving the natural capital in and around agro-ecosystems (Lewis et al., 1997). However, the understanding of IPM varies, so while FAO recently changed its definition of IPM towards greater emphasis on ecologically based management, with pesticides as a last option (W. Settle, presentation to the committee on January 14, 2009 in NRC, 2010), other actors continue to use a narrower definition focusing on improved pesticide use (Shennan et al., 2001).

In pest management, proper soil maintenance to support the microbial, fungal, and nematode community suppressing pathogenic fungi and nematodes, inducing crop resistance responses, and reducing viable weed seed populations is important. Biological control has proven successful for arthropod pest management (NRC, 2010), like the introduction of the parasitoid *Epidinocarsis lopezi* controlling the mealy bug *Phenacoccus manihoti* attacking cassava in Africa (Neuenschwander et al., 2003). Bio-pesticides require that users understand they are working with biological processes or living organisms (Waage, 1996). Thus poor understanding of the function of microbal agents, influenced by the marketing of bio-pesticides as biological versions of conventional pesticides, has reduced the usefulness of many microbial agents (ibid). However, to convince farmers about the value of bio-pesticides and make them choose it over chemical products, the farmers need to be able to assess the impact of the control agent (Williamson, 1997). Conservation biological control enhances indigenous populations of natural enemies such as insects, spiders, and other arthropods by providing habitat in the field (Shennan, 2008), or by planting hedgerows (Letourneau, 1998, Letourneau & Bothwell, 2008, Nicholls et al., 2001). Also "neutral" arthropods, like plankton feeders and detrivores, are important in controlling the pests as they stabilize the natural enemy populations by providing alternative food sources for the latter (Settle et al., 1996).Thus structurally complex landscapes lead to increased parasitism levels and decreased crop damage (Thies & Tscharntke, 1999, Pullaro et al., 2006), as concluded in a review landscape diversity effects on biological control (Tscharntke et al., 2008): "Complex landscapes characterized by highly connected crop-noncrop mosaics may be best for long-term conservation biological control and sustainable crop production".

IPM requires that the farmers, through observation and experimentation, learn about their agro-ecosystem so as to develop site specific technologies and practices (Pretty, 2008). The Farmer Field School (FFS) learning method, which focuses on the farmer as the key decision-maker in pest management and on the facilitation of a discovery-learning process using non-formal education methods (Williamson, 1998), has proven successful in situations of severe pest infestations and excessive synthetic pesticide use (see discussion below). FFS encompasses the following four principles: production of a healthy crop, conservation of natural enemies, performance of regular field observations and belief on the expertise of farmers in their own fields (Pontius et al., 2000). FFS is based on the idea that 'Learning is the process whereby knowledge is created through the transformation of experience' (Kolb, 1984: 38). Experiential learning is a process whereby we, based on the experience of a phenomenon, reflect and allocate meaning to that experience and develop knowledge from it. By experiencing the interactions of the agro-ecosystems and developing their analytical skills, farmers are empowered to realize which factors are within their own control (Fleischer et al., 1999).

According to Long (2001), knowledge is a cognitive and social construction constantly made by the experiences and discontinuities that emerge in the intersection between different actors' 'life worlds', defined as a person's life (or lived) experience, background and values which influence how the person sees the world (Schutz, 1962). Knowledge is a product of dialogue and negotiation, and includes transformation rather than transfer of meaning (Long, 2001). Although farmers may classify observable and culturally important insects better than many entomologists, they may not be aware of the parasitoids and insect pathogens in the agro-ecosystem. This was the case for Honduran smallholders, who distinguished a range of bees and wasps by their flight patterns, but were unaware of the existence of parasitic wasps and the carnivorous nature of wasp larvae (Bentley, 1992). Thus, for the learning and development of knowledge about pest management and the agro-ecosystem complexity, it is important to value both farmers' and scientists' knowledge and experiences. Participatory research approaches combining farmer knowledge and experience with research information could increase the ability to predict when synergies or negative interactions are likely to occur in the field and adjust management accordingly (NRC, 2010). As active farmer participation and public education are often a prerequisite for successful biological control (Williamson, 1998), the lack thereof may reduce the impact of introduced control agents, as was the case during the introduction of the parasitoid

Diadegma semiclausum Hellen (Hym., Ichneumonidae) in Southeast Asia. The parasitoid largely failed to control diamondback moth (*Plutella xylostella* L.) in brassicas in areas where farmers sprayed heavily with broad spectrum insecticides eliminating both native and introduced natural enemies (Waage, 1996, Gyawali, 1997).

IPM-FFS has been used for decades in Asia, where positive social changes (Pontius et al., 2000), have been documented along with reduced pesticide use, increased yield (Pretty & Waibel, 2005) and associated positive human and environmental health effects (e.g. Rola et al., 2001, Erbaugh et al., 2002, Godtland et al., 2004, Praneetvatakul & Waibel, 2006, WB, 2006) such as reduced pest resurgence (Matteson et al., 1994). FFS may provide an arena for shared learning between farmers, scientists and decision makers, with the emphasis on the farmers' experiential learning process, as in the Philippines, Pakistan and Honduras. In a vegetable IPM-FFS in the Philippines, farmers participated in releases of the diamondback moth parasitoid *Diadegma* sp. in their cabbage terraces and built wooden emergence boxes for the parasitized cocoons provided by the local university (ADB, 1996). From their observation of parasitized diamondback moth larvae and exercises demonstrating the effects of commonly used insecticides on the parasitoids, the farmers reduced their pesticide application by 80% and started asking the visiting agrochemical salesmen whether the products they sold were "Diadegma-friendly". Even the mayor of Atok town in the Cordillera region in the Philippines, got convinced about biological control to the extent that he banned all advertising of synthetic insecticides in his municipality (Cimatu, 1997). In Pakistan, FFS was undertaken to tackle the resurgence of the whitefly *Bemisia tabaci* (Gennadius Hom., Aleyrodidae), vectoring cotton leaf curl virus, and the increasing insecticide use in cotton (Poswal & Williamson, 1998). Whitefly may be kept in control by natural enemies, thus the whitefly outbreaks in Pakistan seemed to be a result of the elimination of the key natural enemies in cotton fields due to early and increased insecticide applications. The FFS participants learned about natural enemies and crop compensation by observing whitefly parasitization by Encarsia and Eretmocerus spp. (Hym., Aphelinidae), predation of jassids by mites, ants and spiders, and of whitefly by anthocorid and reduviid bugs, staphylinid beetles and spiders. Due to increased IPM knowledge the participants did not apply any insecticides on the IPM decision making plots the first 8-10 weeks after planting, thus allowing natural enemy populations to build up, and thereby reducing the total number of pesticide applications while yields increased. Having experimented with whitefly resurgence caused by the application of organophosphate, one FFS group demonstrated the impact of unnecessary pesticide applications to the Department of Agriculture officials, neighbouring farmers and local agrochemical salesmen.

In Honduras, the farmers learning about natural enemies and insect reproduction in the Natural Pest Control Course run by Zamorano (the Pan-American School of Agriculture), were able to enhance predation of pests in maize, potatoes and vegetables (Bentley et al., 1994, Rodríguez, 1993). As the farmers had learned the underlying principles of the agro-ecosystem, not merely specific techniques, they could also apply what they had learned to new situations (ibid.). These techniques adapted or invented by the farmers were tailored to their pest problems and resources in a way that standard 'recipes' could never be (ibid.). These results indicate why agricultural systems with high levels of social and human assets are more able to adapt to change and innovate in the face of uncertainty (Uphoff, 1998, Chambers et al., 1989, Bunch & Lopez, 1999, Olsson & Folke, 2001, Pretty & Ward, 2001). However, collective adoption of IPM techniques is vital, because the effect of IPM will be reduced if neighbouring farmers continue relying on chemicals for pest control killing beneficial parasites and predators, and exposing IPM farmers and local ecosystems to chemical spill overs (WB, 2006).

3. Effects of a vegetable IPM-FFS in Cotonou, Benin on pest management

In this section, the vegetable IPM-FFS by IITA-Cotonou will be discussed with emphasis on the extent to which the IPM-FFS training influenced the participants regarding their knowledge and use of IPM options including pesticides, their awareness of health hazards of synthetic pesticides and corresponding handling practices. Also the knowledge created through the interactions between the vegetable producers' and the scientists' 'life worlds' will be explored. The IPM-FFS project conducted two FFS sessions for farmers selected to participate in the Trainer Of Trainers (ToT) in 2003/04, covering the crops *Solanum macrocarpon* L. (Solanaceae) (a variety of the African eggplant, locally known as gboma), *Daucus carota* L. sativus Hayek (Apiaceae) (carrot), *Lactuca sativa* L. (Asteraceae) (lettuce) and *Brassica oleracea* L. capitata L. (Brassicaceae) (cabbage). To scale out the knowledge created in the ToT sessions, each ToT participant was to arrange horizontal sharing of knowledge and skills gained during the ToT sessions. The major factors limiting vegetable production in the urban and peri-urban areas of Benin are soil infertility, pests, weak irrigation infrastructure and poorly developed vegetable enterprises (James et al., 2006). Thus towards improved vegetable pest management, the IITA vegetable project found a strain of the entomopathogenic fungus *Beauveria bassiana* (Balsamo) Vuillemin (Deuteromycotina: Hyphomycetes) (isolate Bba5653) to be effective against the diamondback moth *Plutella xylostella* L., the most devastating pest of cabbage (Godonou et al., 2009a, Godonou et al., 2009b). As the microbial control agent is environmentally friendly and harmless to humans, it was part of the IPM options promoted through FFS training by the project along with botanical nematicides (Loumedjinon et al., 2009), as alternatives to synthetic pesticides.

The FFS training by the IITA vegetable IPM project created interfaces where the knowledge of the vegetable producers was challenged by the knowledge of other producers and scientists. Most of the vegetable producers believed that all arthropods in their fields would damage their crops, and therefore the project focused on distinguishing harmful insects from harmless and beneficial insects (e.g. *Cotesia plutellae* (Kurdjumov) (Hymenoptera: Braconidae), a parasitoid of *P. xylostella*, and the predator Exochomus troberti Mulsant (Coleoptera: Coccinellidae); (James et al., 2006). During the training, the participants were therefore exposed to IPM, beneficial organisms, plant health, agro-ecosystems and the concept of quality vegetables. Agro-ecosystem analysis (AESA) was used to assist participants to base their decisions on whether to apply synthetic pesticides on several criteria connected to observing the development of the plants, pests and diseases in vegetable plots. Vegetable producers were expected to change inappropriate IPM practices as a result of the training offered by the project.

3.1 Methods of data collection

The research was conducted in urban and peri-urban areas of Cotonou, department of Littoral, in 2006/07. In Benin, the law 1991/004 regulates packaging, labelling, transport, storage, usage and disposal of synthetic pesticides on the market, and the national plant protection service (Service de la Protection des Végétaux, SPV) is in charge of the quality control of synthetic pesticides and authorization of salesmen. The Centre d'Action Régional pour le Développement Rural (CADER) was in charge of the control of the use of synthetic pesticides, but it had closed down at the time of the survey. In this study, three vegetable gardens (Houeyiho, Office National d'Edition de Presse et d'Imprimerie (ONEPI) and Gbegamey), where vegetable IPM-FFS had been conducted, were compared with three vegetable gardens in Godomey where no FFS had been held (control group). The gardens where FFS had been held were selected on the basis of accessibility (short travel distance from Cotonou), that they were still in use as vegetable gardens and some of the crops covered in the project (carrot, lettuce, cabbage and gboma) were cultivated there. The control sites were the only vegetable gardens in Cotonou, which had not participated in the project. All the respondents produced vegetables for the market, and nearly all the land area in the gardens studied was used for vegetable production, while the respondents lived in other parts of or outside Cotonou. Each vegetable producer had his/her defined cropping area, consisting of several beds, with an average size of 7.2 m² each (James et al., 2006), but none of them owned the land they cultivated or had formal contracts with the landowner, thus their situation was very insecure (Zossou, 2004).

During the vegetable IPM-FFS, IPM options within the categories 'chemical, biological, mechanical and cultural' were taught. To distinguish how the different vegetable producers understood the concept of IPM, the number of IPM techniques (chemical, biological, mechanical and cultural) in which they mentioned IPM tools when answering the open-ended question, 'What does IPM mean?' was counted. It was also noticed whether the respondents only listed IPM options or explained a holistic approach using AESA. When IPM tools were mentioned within only one IPM approach, the respondents were considered as having a narrow understanding of the concept of IPM. When IPM tools were mentioned within all the four IPM approaches, their understanding of the concept of IPM was considered broad. While the 'concept of IPM' was based on how the respondents described IPM, the respondents' 'knowledge of IPM' was evaluated by the number of IPM techniques, in the four mentioned categories, in which they mentioned IPM tools as response to various questions during the interview. AESA was used in the IPM-FFS training to assist the farmers to take management decisions based on the conditions of their fields. The farmers observe the biotic and abiotic factors, analyse how these impact their crops and thereafter take proper management decisions based on the analysis (Pontius et al., 2000). To perform a sound AESA requires the farmers to have a good understanding of ecosystem interactions such as pest–predator relationships and the existence of beneficial insects. Insect zoos (enclosed pot with a plant, a pest and/or a beneficial insect) are often used to visualize the pest–predator relationships (Pontius et al., 2000).

A transect walk was done in all the vegetable gardens to get preliminary information about the area. Convenience sampling of the snowball type was used (Bryman, 2004), and the most available vegetable producers in the gardens were identified. Among these, the sample of producers was selected on the basis of gender, age, education, and economic and social status. The list of respondents was checked by the leadership of the community-based farmers union in Cotonou (Union Communale des Producteurs, UCP) to ensure that all socioeconomic categories were represented. From the IPM-FFS project area, 15 ToT participants, 9 FFS participants and 19 non-participants were selected. Twelve control respondents were selected from the area where no IPM-FFS had been conducted. Fifty-four semi-structured interviews with open-ended questions were carried out with the vegetable producers. The interviews focused on knowledge and use of IPM options, awareness of health hazards of synthetic pesticides, handling practices of synthetic pesticides, and knowledge and use of protection gear. Focus group interviews were held with the producers in the ToT, FFS and non-participant groups to collect data on synthetic and botanical pesticides, the IPM-FFS training and the production environment in the vegetable gardens. Female and male producers were interviewed separately for open discussions. Key informant interviews were held with an ambulant salesman of agro-chemical inputs; two elderly, experienced vegetable producers; and various NGO staff and government employees, as well as with project stakeholders from an NGO specialized in biological agriculture (Organisation Béninoise pour la Promotion de l'Agriculture Biologique), SPV, the national institute of agricultural research in Benin (Institut National de Recherche Agricole du Bénin), IITA in Benin and the UCP. Data was collected on pesticides (rules, regulations and sales) and on the IPM-FFS training (structure and curriculum). Triangulation and follow-up questions were used within the interviews to capture the respondents' real view. An interpreter was used for all the interviews with the vegetable producers so that the interviews were held in the common local language 'Fon' (spoken by most of the interviewed vegetable producers) or French.

As far as possible, the double difference model, comparing the differences in change over time between two populations, was used. The change in behaviour within the ToT and FFS groups was compared with the change in behaviour of the non-participants and control groups. A modified version of Mangan and Mangan's (Mangan & Mangan, 1998) model was used to assess producers' understanding of beneficial insects. The questions 'If you had the possibility would you like to kill all the insects in your field?' and 'Are there any insects that might be beneficial to have in your field (if yes give an example)?' were asked at different points during the interview. If the respondent did not want to kill all insects and could give examples of beneficial insects, she/he would be grouped as having a 'good concept' of beneficial insects.

3.2 Results and discussion
3.2.1 Increased IPM knowledge and plant health

All the ToT respondents, 56% of the FFS respondents, but only 16% of the non-participants were familiar with the term IPM. The ToT respondents had a broader understanding of the concept of IPM, but in general, most of the vegetable producers who were familiar with IPM had a narrow understanding of the concept of IPM (Fig. 1).

All the producers understood IPM as a separate management tool, but while most of them associated IPM with chemical control, the ToT and FFS respondents were more concerned about reducing the use of synthetic pesticides and using botanicals (such as neem, papaya, pepper, orange and cassava epidermis). 'Knowledge of IPM' was based on the IPM tools reported by the respondents during the interview. All the respondents had a broader 'knowledge of IPM' than 'concept of IPM', indicating that even though many vegetable producers were not familiar with the scientific term 'IPM', they knew about various pest management methods. A larger proportion of the ToT respondents (60%) had a good concept of beneficial insects than the FFS (22%), control (18%) and non-participant respondents (none). As the ToT participants were only shown pictures of the beneficial and harmful insects, but did not have any insect zoo, they lacked the experience of visually understanding pest–natural enemy relationships. As a consequence, many did not transform the information into reliable knowledge, as illustrated by a ToT participant: "In the ToT I learned that beneficial insects eat the pest, but it is too risky to rely on it so I rather kill all the insects. I have to see how the beneficial insects behave in practice before I can trust that they won't damage my vegetables, but I don't have enough space to experiment with this". In terms of AESA activities, the largest improvements in observing pests and crop interactions were among the ToT respondents (Fig. 2), who shifted from using preventive application to applications based on frequent observations of changes in crop, pest and natural enemy developments.

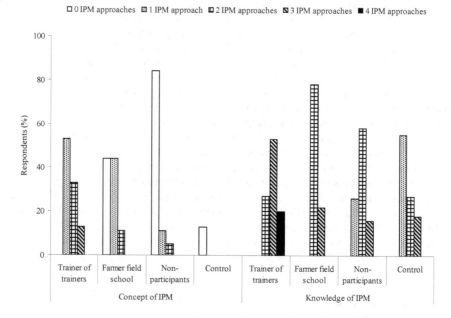

Fig. 1. Broadness of "Concept of IPM" and "Knowledge of IPM"

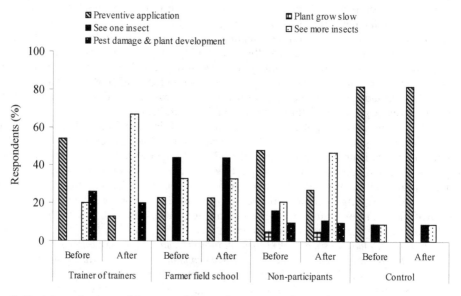

Fig. 2. Decision criteria used by vegetable producers to apply synthetic pesticides before and after the training

Sixty-seven percent of the FFS respondents said that the project had reinforced their knowledge to observe pests and crops, as told by a ToT participant: "*Now (after the Project) I greet my plants in morning and ask how they spent the night. If bad, I have to consider what to do*". Nevertheless, only one ToT respondent used agro-ecosystem analysis and based the management decisions on a holistic analysis of the field; the rest based their decisions on subjective assessment of increasing pest densities. The rate of calendar application was higher when the producers did not have a good understanding of beneficial insects, indicating that having a good understanding of beneficial insects may reduce the calendar application or vice versa. The largest change was among the ToT respondents (33%), shifting from consulting family and neighbours and not measuring at all, to reading the label and using information from the ToT/FFS training.

The term 'plant health', seemed to be a well-known term in the area, but how the respondents understood plant health differed. While the majority of the FFS (89%), non-participant (68%) and control respondents (63%) emphasized the use of synthetic pesticides to get healthy plants, the ToT respondents placed more importance on botanically and biologically based pesticides. Using organic matter to build the soil is important for plant health, and the ToT and FFS respondents were more aware of the importance of applying organic matter before sowing. However, the use of compost seemed to be dependent on the availability and price, as 82% of the respondents in Houeyiho used it, but nearly none of the respondents from the other areas used it. Also seed quality is important for plant health, and only ToT respondents (67%) knew how to use germination testing to check the seed quality, but even though all but one of them claimed to use it, no beds where germination testing was performed were observed during this study. The experiments in the project consisted of giving two beds different treatments and comparing the results. None of the respondents in the project experimented in this way as a result of the training. Forty-seven per cent of the ToT respondents and 22% of the FFS respondents, however, experimented in other ways after the ToT/FFS training, meaning that they tried out methods taught in the project such as the dose of synthetic pesticides, botanical pesticides, observing their fields, and organic and chemical fertilizers. While one respondent said: "*Cultivating vegetables is about experimenting*", others were more risk averse, and lack of time and land constraints were the most common reasons for not experimenting.

The respondents used 32 different types of synthetic pesticides (Fig. 3) and of these pesticides all the class 1b, but also some class II (Endosulfan and Fenpropathrin) and even class III pesticides (Orthene and Malathion) contained substances that are banned or severely restricted in the European Union (PANEurope, 2009), posing serious health concerns for the producers, consumers and the environment. Two respondents used endosulfan, which is prohibited in Benin and proposed by the POPs Review Committee to be eliminated from the global market (StockholmConvention, 2010).

ToT respondents were more often applying the correct pesticides against targeted pests than the other respondents. On the other hand, a large proportion of the control (82%), ToT (80%), FFS (44%) and non-participant respondents (42%) used 'cocktails', mixing up to four different pesticides. Two respondents illustrated the producers' difficult situation: "*I use Talstar against leaf miners and field crickets although I know it is not effective against those pests*" and "*I have no solution for the nematodes on my gboma. I think Kinikini would be effective, but I do not have money to buy it*". The use of cocktails of different pesticides made it difficult to evaluate the project's impact on frequency and quantity of pesticides used (Table 1).

The project recommended to use correct pesticides and to follow the prescriptions from the manufacturer, but as one respondent said: "*The pesticides they (the Project) recommended us to use are not available, so we have to use what is available*". Thus, due to unavailability and expensiveness, the majority of the producers sometimes or always bought synthetic pesticides in non-original packages not knowing what they were using. Even if the vegetable producers bought synthetic pesticides in the original packages, some respondents might be illiterate and the labels were often in foreign languages. Most of the respondents were very much aware of and respected the recommended pre-harvest interval. As indicated by one respondent, economic constraints may reduce the safety in pesticide applications: "*There is shortage of land and people need money so it is difficult to wait the recommended days before harvest*". On the other hand, the safety is also influenced by the individual and collective perception of ethics, as one respondent said: "*It has to be made socially unacceptable to not respect the pre-harvest interval*" and another respondent describing the changes due to increased awareness: "*Earlier some people sold the cucumber five days after spraying, but now everybody know they have to respect the harvest interval, so people will inform the buyer if the vegetable producer has sprayed too close to the harvest*".

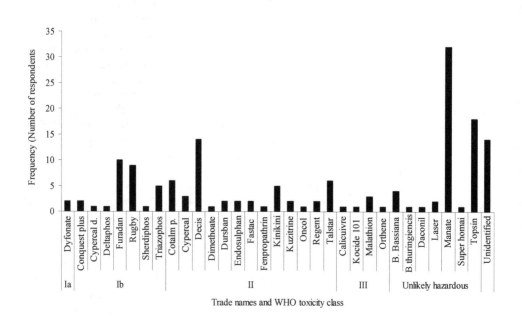

Fig. 3. Pesticides used by the respondents

Synthetic pesticides	Vegetable	ToT	FFS	NP	C	PB	ToT	FFS	NP	C	PB
		Frequency					Quantity				
		(times/crop season)					(l or g/ha)				
Decis (l)	Cabbage	27	-	33	10	7	19	-	138	46	31,9
	Gboma	2	2	3	3	5	2	2	-	6	7
	Lettuce	2	1	4	3	-	1	-	10	8	-
	Carrot	3	-	-	-	1	3	-	-	-	3,5
Talstar (l)	Cabbage	3	6	-	-	12	5	13	-	-	14,91
	Gboma	2	2	-	3	2	3	2	-	6	7
	Lettuce	2	1	4	3	2	1	-	10	8	7
	Carrot	1	-	-	-	-	3	-	-	-	-
Manate (g)	Cabbage	11	-	-	-	-	6	-	-	-	-
	Gboma	2	3	2	3	2	4	30	7	10	13,9
	Lettuce	2	2	2	3	2	3	3	4	13	19,9
	Carrot	-	-	-	-	-	-	-	-	-	-
Fenpropathrin (g)	Cabbage	-	-	18	-	-	-	-	9	-	-
	Gboma	-	-	-	-	-	-	-	-	-	-
	Lettuce	-	-	3	-	-	-	-	5	-	-
	Carrot	-	-	-	-	-	-	-	-	-	-
Endosulfan (l)	Cabbage	1	-	-	-	-	0,4	-	-	-	-
	Gboma	1	-	-	-	-	1	-	-	-	-
	Lettuce	1	2	-	-	-	1	6	-	-	-
	Carrot	1	-	-	-	-	3	-	-	-	-
Furadan (g)	Cabbage	-	-	-	-	-	-	-	-	-	-
	Gboma	-	1	-	-	3	-	46	-	-	13,9
	Lettuce	1	-	-	-	-	46	-	-	-	-
	Carrot	1	1	1	-	2	-	-	-	-	13,9

Table 1. Average frequency and quantity[1] of synthetic pesticides per growing season

[1]based on the quantity of concentrated synthetic pesticide, not quantity of active ingredients. NP = Non-participants, C = Control, PB = Project baseline data

The project emphasized the use of botanical pesticides as an alternative to synthetic pesticides. Forty-seven per cent of the ToT respondents, 26% of the non-participants, 22% of the FFS respondents and 9% of the control respondents said that they used neem, but during the time of this survey, no vegetable producers were observed preparing botanicals. The reason for not using neem extract was the time-consuming and labour intensive preparation as one needs large quantities of leaves or fruits. If they could buy neem extract commercially, many may use it because they had experience with it and appreciated that it was not harmful to the environment or to the individuals who applied it. In the project, knowledge about biological control was introduced. B. bassiana was used to control diamondback moth, and the vegetable producers could request it from IITA. Many of the vegetable producers were pleased with B. bassiana as it saved them money and labour, and their interest in the product was expressed by a ToT participant: *"When you use bassiana (B. bassiana) you are sure to succeed in growing cabbage. If I didn't have bassiana I would have stopped growing cabbage"*. At the time, B. bassiana was given for free, but if the product is commercialized, the question remains whether the price will be competitive with respect to synthetic pesticides so the producers can afford it. However, increasing evidence of resistance in the diamondback moth may force producers to use alternatives or to abandon cabbage as a crop. Crop rotation was traditional knowledge used by all the producers, but they became more aware of its importance because of the ToT/FFS training. Even if some vegetable producers were not able to explain in scientific terms what was happening in their crops, they improved their practices based on experience, like one control respondent said: *"Normally you only have to apply pesticides three times, but if you grow the same type of vegetable two times after each other you need five pesticide applications"*. All the ToT respondents, 94% of the non-participants, 78% of the FFS respondents and 73% of the control respondents practiced intercropping, but the main reasons for this practice were economic gains and land shortage. Also most of the respondents chose the planting time for economic motives considering market prices, while none considered plant health issues.

3.2.2 Awareness of health hazards from synthetic pesticides and proper handling practices

Awareness of negative effects of synthetic pesticides was quite high among the respondents, with the most known effect being hazards to the farmers' health as one respondent noted: *"It is not good to apply pesticides as it makes my eyes burn"*. The control group had a more limited view of the negative effects of synthetic pesticides, and only mentioned human, consumer and farmer health, but not environmental and long term effects.

The overall awareness about protection while spraying synthetic pesticides was high (Fig. 4), as one respondent said: *"You should cover all the parts of your body. You may not feel anything today, but the pesticides will accumulate in your body and in some years cause heart problems and headache"*, but the proportion actually using such equipment was low. Among the control respondents, 46% only wore shorts and T-shirts while spraying synthetic pesticides, but virtually none of the other respondents could show any of their protection gear. The most common reason for not using any protection device was expense, while another common reason in the tropics is the heat. The most common post-spray activity among the producers was to take a bath, done by most of the ToT respondents, while the control respondents were more likely to only wash their hands, legs and face. Most of the vegetable producers stored their synthetic pesticides buried in the vegetable field, hid it in the bushes nearby, or at home. Only the ToT respondents from ONEPI stored synthetic pesticides in a storage room, which could be locked, as they had a common storage room in the garden. Among the respondents buying original pesticide packages, all the FFS and control respondents, 62% of the non-participants and 22% of the ToT respondents sometimes stored the synthetic pesticides in empty soft drink bottles. When using pesticide bottles without labels or storing it in soft drink bottles, other people in the household may use the content in the belief that it is something else. Many threw the pesticide cans in the garbage or bushes where pesticide may leach out in the ground or accidents may happen if children find them. The most common way to apply pesticides was to use a spray sack, but 27% of the control respondents, mixed the synthetic pesticides with water in a bowl and used a bunch of grass to 'paint' the vegetables.

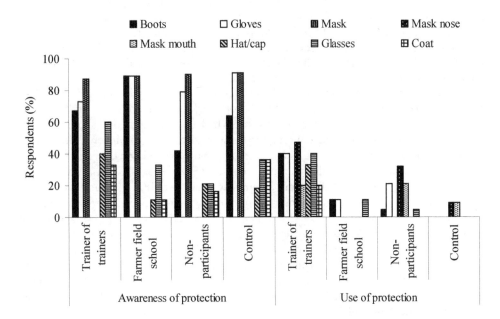

Fig. 4. Awareness and use of protection during application of synthetic pesticides

To scale out IPM knowledge and practices, the ToT participants were expected to carry out FFS sessions on all the four vegetables in the project. However, the FFS respondents' general complaints were that (1) the FFS sessions were often not conducted in both seasons and did not include all the four crops in the project and (2) the quality of the training was lower than that of the ToT sessions. The FFS sessions started the same season as the ToT sessions, thus not allowing the ToT respondents' time to develop their facilitation skills in IPM before commencing as trainers. Initially, the project only had IPM plots, but had no plots where the farmers' existing practice was demonstrated, which indicated a poor emphasis of the project on the vegetable producers' own experimentation and knowledge-creating processes.

4. Conclusion

The project increased the ToT participants' knowledge about IPM, reflected in their good concepts of beneficial insects, improved knowledge about plant health and pest management tools, improved ability to take management decisions based on pest occurrence in the field and increased experimentation with knowledge gained from the project. Increased knowledge and awareness about IPM may be one of the reasons for the participants' change in attitude towards synthetic pesticides. While some claimed having adopted biological IPM tools, the participants had not changed their practices significantly regarding the use of synthetic pesticides and cocktails, and most of the producers did not apply correct pesticides to the target pests. The FFS participants had gained considerably less knowledge than the ToT participants, which may be due to less intensive FFS sessions led by trainers with less developed facilitating skills. Experiential learning reflects and allocates meaning to that experience and develops knowledge from it. There were several examples of this issue in the project area, where vegetable producers, who did not know scientific terms such as IPM or nematodes, had nevertheless learned IPM practices based on experience. In the project, however, access to information about beneficial insects was mainly through theory, thus the participants did not experience what 'beneficial insect' meant in practice, and consequently did not transform this information into meaningful knowledge.

The practical use of concepts such as 'IPM' and 'agro-ecosystem' requires an understanding of complex interactions, which takes time to develop. Lack of monitoring of activities in their own fields is one reason why very few participants had a holistic and good understanding of those concepts. While it remains important to bring in scientific information to improve the vegetable producers' understanding, this information should be built on the participants' local knowledge to make sense to them and be relevant. The results show that the surveyed participants in the project did not adopt a complete package of IPM tools and concepts, but rather experimented with the new information and thereafter adapted parts of what they learned into their production systems. While Mancini (2006) found a strong correlation between knowledge level and the reduction in pesticide use among Indian farmers attending cotton IPM-FFS, also other studies show that farmers with a good understanding of how the field ecosystem works perform better crop management than those who get discrete and simplified pest management instructions (Mangan & Mangan, 1998, Price, 2001). However, in Benin, even the project participants with a good understanding of beneficial insects used superfluous pesticides, which might kill the beneficial insects. This shows that it requires more than a good understanding to change usual practices.

There are many factors in the vegetable producers' environment hindering them in using IPM and using synthetic pesticides more safely, including lack of ecological knowledge and access to product information, little availability of the right products and their relatively high cost. Many participants wished to produce good quality vegetables with safer use of synthetic pesticides. However, limited access to land and wish to make higher profits, were strong driving forces leading some vegetable producers to unsafe practices, such as not respecting the harvest interval and using forbidden synthetic pesticides. Also, normative considerations influence the vegetable producers' practices. As the laws on pesticide use are not enforced in Benin, these have limited impact on people's behaviour, but as seen from the survey, the awareness rising from various NGOs and research institutions has changed the vegetable producers' attitudes towards respecting the pre-harvest interval. Even more awareness rising is needed to change the producers' attitude towards synthetic pesticides and make dangerous practices unacceptable. Maybe even analysis of pesticide residues of products from individual beds to personalize the information could be an input that is an asset for producers to fully understand the problem of residues in the products.

The results indicate possible trade-offs between health and economic effects, with the latter weighing more heavily. The awareness of pesticide hazards and proper protection gear was generally high, and although many producers had experienced negative health effects of synthetic pesticides, most producers still did not use protection gear due to expense. The project did not have any impact on the safety in storage and handling of pesticides as the practices were rather influenced by what was more convenient for the vegetable producers.

This study shows that there is a need to focus more on the vegetable producers' own knowledge creation by emphasizing experiential learning, as well as to enable the producers to realize their role as potential knowledge generators (Simpson & Owens, 2002). This is in line with other studies emphasizing the importance of establishing a dynamic process where the participants take control over the experimentation (Braun et al., 2004) and on developing a 'learning style' (Pretty, 1995) that enables 'exploration, evaluation and adaptation of technological alternatives' (Lee, 2005 : 1332). Post-IPM-FFS activities would probably allow the participants in the Cotonou vegetable IPM-FFS to develop their ideas and concepts about IPM and agro-ecosystems as they practice these in their fields. There were many competent and knowledgeable vegetable producers in the studied areas concerned about the dangerous use of pesticides. In a follow-up of the IPM-FFS activities, these people could be the driving forces of promoting and implementing IPM in their communities. The general results from this study in Benin are in line with other reports (Maumbe et al., 2003, Mancini, 2006), concluding that for IPM to succeed as a proper and reliable plant protection strategy, not only is there a need to consider the educational component involving individual farmers or groups of farmers, but it is also necessary for all the stakeholders involved (farmers, extension, scientists, policy makers and NGOs) to understand the complex nature of IPM. In addition to the educational component, all other factors, such as the need for group versus individual action, farmer's indigenous knowledge, farmer's resource endowments, and last but not least the macroeconomic determinants, do play a significant role in establishing whether IPM can succeed or not. In light of findings from Benin and other IPM-FFS programs, further research is needed on how to facilitate the processes of knowledge creation between the farmers and scientists, and how to involve the ToT participants' in a way that they feel the commitment to continue the learning processes, to share knowledge with their farming communities and to initiate changes in pest management at the community level.

5. Acknowledgement

Thanks to all who contributed with their experience and knowledge to this research.
This chapter has drawn upon material from 'T. Lund, M.-G. Sæthre, I. Nyborg, O. Coulibaly and M. H. Rahman, "Farmer field school IPM impacts on urban and peri-urban vegetable producers in Cotonou, Benin", **International Journal of Tropical Insect Science**, 2 Volume 30 (1), pp 19-31, (2010) © ICIPE, Published by Cambridge University Press, reproduced with permission'.

6. References

ADB. 1996. *Integrated pest management for highland vegetables.* Asian Development Bank, Manila, Philippines

Akogbeto, M. C., Djouaka, R. & Noukpo, H. 2008. Utilisation des insecticides agricoles au Bénin. *Entomologie médicale. BMC Genomics,* 9, 1–10

Arcury, T. A., Quandt, S. A. & Mellen, B. G. 2003. An exploratory analysis of occupational skin disease among Latino migrant and seasonal farmworkers in North Carolina. *J Agric Saf Health,* 9, 3, pp. 221–232

Ascherio, A., Chen, H., Weisskopf, M. G., O'Reilly, E., McCullough, M. L., Calle, E. E., Schwarzschild, M. A. & Thun, M. J. 2006. Pesticide exposure and risk for Parkinson's disease. *Annals of Neurology,* 6, 2, pp. 197–203

Atcha-Ahowé, C., James, B., Godonou, I., Boulga, J., Agbotse, S. K., Kone, D., Kogo, A., Salawu, R. & Glitho, I. A. 2009. Status of chemical control applications for vegetable production in Benin, Ghana, Mali, Niger, Nigeria and Togo – West Africa. *Pesticides Management in West Africa,* 7, pp. 4–14

Awmack, C. S. & Leather, S. R. 2002. Host plant quality and fecundity in herbivorous insects. *Annual Review of Entomology,* 47, 817–844

Bawden, R. 2005. *The Hawkesbury experience: tales from a road less travelled,* Earthscan, London

Beanland, L., Phelan, P. L. & Salminen, S. 2003. Micronutrient interactions on soybean growth and the developmental performance of three insect herbivores. *Environmental Entomology,* 32, 3, pp. 641–651

Bentley, J. W. 1992. Alternatives to pesticides in central America; applied studies of local knowledge. *Culture & Agriculture,* 44, 10-13

Bentley, J. W., Boa, E., van Mele, P., Almanza, J., Vasquez, D. & Eguino, S. 2003. Going public: a new extension method. *Int. J. Agric. Sustainability,* 2, 108–123

Bentley, J. W., Rodriguez, G. & Gonzalez, A. 1994. Science and people: Honduran campesinos and natural pest control inventions. *Agriculture & Human Values,* 11, 178-183

Beseler, C. L., Stallones, L., Hoppin, J. A., Alavanja, M. C. R., Blair, A., Keefe, T. & Kamel, F. 2008. Depression and pesticide exposures among private pesticide applicators enrolled in the Agricultural Health Study. In: *Environ. Health Perspect,* 30.12.2010, Available from:
http://www.pubmedcentral.nih.gov/articlerender.fcgi?tool=pmcentrez&artid=259 9768

Braun, A., Herrera, I., Rosset, P., Teurings, J. & Vedeld, P. 2004. *Ecologically-based participatory implementation of integrated pest management and agroforestry in Nicaragua and Central America (CATIE-IPM/AF) Phase III.*

Bruce, T. J. & Pickett, J. A. 2007. Plant defence signalling induced by biotic attacks. *Current Opinion in Plant Biology,* 10, 4, pp. 387–392

Bryman, A. 2004. *Social research methods,* Oxford University Press Inc., New York

Bunch, R. & Lopez, G. 1999. Soil recuperation in Central America. In: *Fertile Ground: The Impact of Participatory Watershed Management,* Hinchcliffe, F., Thompson, J., Pretty, J. N., Guijt, I. & Shah, P. (eds.), pp. 32-41, Intermediate Technology Publ, London

Chambers, R. 2005. *Ideas for development,* Earthscan, London

Chambers, R., Pacey, A. & Thrupp, L. A. (eds.). 1989. *Farmer First: Farmer Innovation and Agricultural Research,* IT Publications, London.

Cimatu, F. 1997. War on pesticide adverts. *Pesticides News,* 38, 21

Cole, D. C., Sherwood, S., Crissman, C., Barrera, V. & Espinosa, P. 2002. Pesticides and health in highland Ecuadorian potato production: assessing impacts and developing responses. *International Journal of Occupational & Environmental Health,* 8, 3, pp. 182-190

Dent, D. 1995. *Integrated Pest Management,* Chapman & Hall, London

Djouaka, R. F., Bakare, A. A., Coulibaly, O. N., Akogbeto, M. C., Ranson, H., Hemingway, J. & C., S. 2005. Expression of the cytochrome P450s CYP6P3 and CYP6M2 are significantly elevated in multiple pyrethroid resistant populations of Anopheles gambiae s.s. from Southern Benin and Nigeria. *Bulletin of the Exotic Pathology Society,* 98, 400–405

Erbaugh, J. M., Donnermeyer, J., Kibwika, P. & Kyamanywa, S. 2002. An assessment of the integrated pest management collaborative research support project's (IPM CRSP) activities in Uganda: Impact on farmers' awareness and knowledge of IPM skills. *African Crop Science Journal,* 10, 3, pp. 271 - 280

FAO & ILO. 2009. *Safety and Health,* Food and Agriculture Organization and International Labour Organization, Available from: http://www.fao-ilo.org/fao-ilo-safety/en/

Fleischer, G., Jungbluth, F., Waibel, H. & Zadoks, J. C. 1999. *A field practioner's guide to economic evaluation of IPM,* Uni Druck Hannover, Hannover

Folke, C. 2006. Resilience: the emergence of a perspective for social-ecological systems analyses. *Global Environmental Change,* 16, 253–267

Fox, J. E., Gulledge, J., Engelhaupt, E., Burow, M. E. & McLachlan, J. A. 2007. Pesticides reduce symbiotic efficiency of nitrogen-fixing rhizobia and host plants. *Proceedings of the National Academy of Sciences of the United States of America,* 104 24, pp. 10282-10287

Gallagher, K., Ooi, P., Mew, T., Borromeo, E., Kenmore, P. & Ketelaar, J.-W. 2005. Ecological basis for low-toxicity integrated pest management (IPM) in rice and vegetables. In: *The pesticide detox,* Pretty, J. (ed.), pp. Earthscan, London

Godonou, I., James, B. & Atcha-Ahowé, C. 2009a. Locally available mycoinsecticide alternatives to chemical pesticides against leaf feeding pests of vegetables. *Pesticides Management in West Africa,* 7, pp. 53–62

Godonou, I., James, B., Atcha-Ahowé, C., Vodouché, S., Kooyman, C., Ahanchedé, A. & Korie, S. 2009b. Potential of *Beauveria bassiana* and *Metharhizium anisopliae* isolates from Benin to control *Plutella xylostella* L. (Lepidotera: Plutellidae). *Crop Protection,* 28, 220–224

Godtland, E. M., Sadoulet, E., Janvry, A. d., Murgai, R. & Ortiz, O. 2004. *The impact of farmer field schools on knowledge and productivity: A study of potato farmers in the Peruvian Andes.* University of Chicago, Chicago

Gyawali, B. K. Year. Farmer field schools: a participatory outreach research approach on safe application of agro-chemicals and bioproducts. In: Salokhe, V. M., ed. Proceedings of the international workshop on safe and efficient application of agrochemicals and bio-products in South and Southeast Asia, 1997 Bangkok, Thailand. Asian Institute of Technology.

Holling, C. S., Berkes, F. & Folke, P. 1998. *Linking social and ecological systems: management practices and social mechanisms for building resilience,* Cambridge University Press, Cambridge

James, B., Godonou, I., Atcha, C., Baimey, H., Adango, E., Boulga, J. & Goudegnon, E. 2006. *Healthy vegetables through participatory IPM in peri-urban areas of Benin.* IITA, Cotonou

Kishi, M. 2005. The Health Impacts of Pesticides: What do we now know? In: *The Pesticide Detox,* Pretty, J. (ed.), pp. Earthscan, London

Kolb, D. 1984. *Experiential Learning,* Prentice Hall, Inc, New Jersey

Lee, D. R. 2005 Agricultural sustainability and technology adoption: Issues and policies for developing countries. *American Journal of Agricultural Economics,* 87, 5, pp. 1325–1334

Letourneau, D. K. 1998. Conserving biology, lessons for conserving natural enemies. In: *Conservation Biological Control,* Barbosa, P. (ed.), pp. Academic Press., San Diego, California

Letourneau, D. K. & Bothwell, S. G. 2008. Comparison of organic and conventional farms: challenging ecologists to make biodiversity functional. *Frontiers in Ecology and the Environment,* 6, 8, pp. 430–438

Lewis, W. J., van Lenteren, J. C., Phatak, S. C. & Tumlinson, J. H. 1997. A total system approach to sustainable pest managmement. *Proc. Natl Acad. Sci. USA,* 94, 12243–12248

Lichtenberg, E. 1992. Alternative approaches to pesticide regulation. *Northeast J. Agric. and Resource Economics,* 21, 83-92

Liebelin, G., Østergaard, E. & Francis, C. 2004. Becoming an agroecologist through action education. *International Journal of Agricultural Sustainability,* 2, 147–153

Lim, G. S. 1992. Integrated pest management in the Asia-Pacific context. In: *Integrated pest management in the Asia-Pacific region. Proceedings of the conference on integrated pest management in the Asia-Pacific region, 23-27 September 1991, Kuala Lumpur, Malaysia,* pp. 459, CAB International, Wallingford, UK

Long, N. 2001. *Development sociology actor perspectives,* Routledge, London

Loumedjinon, S., Baimey, H. & James, B. 2009. Locally available botanical alternatives to chemical pesticides against root-knot nematode pests of carrot (*Daucus carota*) in Benin. *Pesticides Management in West Africa,* 7, 34–52

Lund, T., Sæthre, M.-G., Nyborg, I., Coulibaly, O. & Rahman, M. H. 2010. Farmer field school IPM impacts on urban and peri-urban vegetable producers in Cotonou, Benin. *International Journal of Tropical Insect Science,* 30, 1, pp. 19-31

Mancini, F. 2006. *Impact of integrated pest management farmer field schools on health, farming systems, the environment, and livelihoods of cotton growers in Southern India.* PhD, Wageningen University.

Mancini, F., Van Bruggen, A. H. C., Jiggins, J. L. S., Ambatipudi, A. C. & Murphy, H. 2005. Acute pesticide poisoning among female and male cotton growers in India. *International Journal of Occupational and Environmental Health,* 11, 221-232

Mangan, J. & Mangan, M. S. 1998. A comparison of two IPM training strategies in China: The importance of concepts of the rice ecosystem for sustainable insect pest management. *Agriculture and Human Values*, 15, 3, pp. 209–221

Matteson, P. C., Gallagher, K. D. & Kenmore, P. E. 1994. Extension ofintegrate d pest management for plant hoppers in Asian irrigated rice: empowering the user. In: *Ecology and management of planthoppers*, Denno, R. F. & Perfect, T. J. (eds.), pp. 656-685, Chapman & Hall, London

Maumbe, B., Bernstein, R. & Northon, G. 2003. Social and economic considerations in the design and implementation of integrated pest management in developing countries. In: *Integrated pest management in the global arena*, Maredia, K., Dakouo, D. & Mota-Sanchez, D. (eds.), pp. 87-95, CAB International,

McIntyre, B. D., Herren, H. R., Wakhungu, J. & Watson, R. T. (eds.). 2009. *Agriculture at a Crossroads*, International Assessment of Agricultural Knowledge, Science and Technology for Development. A Synthesis of the Global and Sub-Global IAASTD Reports, Washington.

Miller, G. T. & Spoolman, S. 2009. *Sustaining the Earth: an integrated approach*, Thompson Learning, Inc., Pacific Grove, California

Neuenschwander, P., Borgemeister, C. & Langewald, J. (eds.). 2003. *Biological Control in IPM Systems in Africa*, CAB International, Wallingford, UK.

Nicholls, C. I., Parrella, M. & Altieri, M. A. 2001. The effects of a vegetational corridor on the abundance and dispersal of insect biodiversity within a northern California organic vineyard. *Landscape Ecology*, 16, 2, pp. 133–146

Norse, D., Ji, L., Leshan, J. & Zheng, Z. 2001. *Environmental costs of rice production in China*, Aileen Press, Bethesda, MD

NRC. 1999. *Our Common Journey: Transition towards sustainability*, National Academy Press, Washington DC

NRC. 2010. *Toward Sustainable Agricultural Systems in the 21st Century*, Committee on Twenty-First Century Systems Agriculture. National Research Council, National Academies Press,

O'Malley, M. A. 1997. Skin reactions to pesticides. *Occupational Medicine*, 12, 2, pp. 327–345

Olsson, P. & Folke, P. 2001. Local ecological knowledge and institutional dynamics for ecosystem management: a study of Lake Racken watershed, Sweden. *Ecosystems*, 4, 85-104

Orr, D. 1992. *Ecological literacy*, SUNY Press, Albany, NY

PANEurope. 2009. *What substances are banned and authorised in the EU market ?* , Pesticide Action Network Europe, 05.01.2011, Available from: http://www.pan-europe.info/Archive/Banned%20and%20authorised.htm

Pimbert, M., Barry, B., Berson, A. & Tran-Thanh, K. 2010. *Democratising Agricultural Research for Food Sovereignty in West Africa*. IIED, CNOP, Centre Djoliba, IRPAD, Kene Conseils, URTEL, Bamako and London

Pingali, P. L. & Roger, P. A. 1995. *Impact of pesticides on farmers' health and the rice environment*, Kluwer, Dordrecht, The Netherlands

Pontius, J., Dilts, R. & Bartlett, A. 2000. *From Farmer Field Schools to Community IPM: Ten years of building community*, FAO Community IPM programme, Jakarta

Poswal, A. & Williamson, S. 1998. *Stepping off the cotton pesticide treadmill: preliminary findings from a farmer participatory cotton IPM training project in Pakistan.* Ascot, UK

Praneetvatakul, S. & Waibel, H. 2006. *Farm Level and Environmental Impacts of Farmer Field Schools in Thailand.* Development and Agricultural Economics, Faculty of Economics and Management, University of Hannover, Germany,

Pretty, J. 2008. Agricultural sustainability: concepts, principles and evidence. *Philosohical Transactions of The Royal Society Biological Sciences,* 363, 1491, pp. 447-465

Pretty, J. & Waibel, H. 2005. Paying the price: the full cost of pesticides. In: *The pesticide detox,* Pretty, J. (ed.), pp. Earthscan, London

Pretty, J. N. 1995. *Regenerating agriculture,* earthscan Publications Limited, London, UK

Pretty, J. N. & Ward, H. 2001. Social capital and the environment. *World Development,* 29, 1, pp. 209-227

Price, L. L. 2001. Demystifying farmers entomological and pest management knowledge: A methodology for assessing the impacts on knowledge from IPM-FFS and NES interventions. *Agricultural and human values,* 18, 2, pp. 153 - 156

Pullaro, T. C., Marino, P. C., Jackson, D. M., Harrison, H. F. & Keinath, A. P. 2006. Effects of killed cover crop mulch on weeds, weed seeds, and herbivores. *Agriculture Ecosystems & Environment,* 115, 1-4, pp. 97–104

Rodríguez, G. 1993. *Experimentación y Generación de Tecnologías en Control Natural de Plagas con Pequeños Agricultores en Honduras.* Ingeniero Agrónomo Thesis, Escuela Agrícola Panamericana.

Rola, A. & Pingali, P. 1993. *Pesticides, rice productivity, and farmers' health an economic assessment.* IRRI, Los Baños, The Philippines

Rola, A. C., Quizon, J. B. & Jamias, S. B. 2001. Do Farmer Field School Graduates Retain and Share What They Learn?: An Investigation in Iloilo, Philippines. *Journal of International Agricultural and Extension Education,* 9, 1, pp. 65-76

Rosendahl, I., Laabs, V., Atcha-Ahowé, C., James, B. & Amelung, J. 2009. Insecticide dissipation from soil and plant surfaces in tropical horticulture of southern Benin,West Africa. *Journal of Environmental Monitoring,* 11, 1157–1164

Rosenthal, E. 2005. Who's afraid of national laws? Pesticide corporations use trade negotiations to avoid bans and undercut public health protections in Central America. *International Journal of Occupational and Environmental Health,* 11, 437-443

Schutz, A. (ed.) 1962. *The Problem of Social Reality,* Kluwer, Dordrecht.

Settle, W. H., Ariawan, H., Astuti, E. T., Cahyana, W., Hakim, A. L., Hindayana, D., Lestari, A. S. & Pajarningsih. 1996. Managing tropical rice pests through conservation of generalist natural enemies and alternative prey. *Ecology,* 77, 7, pp. 1975–1988

Shennan, C. 2008. Biotic interactions, ecological knowledge and agriculture. *Philosophical Transactions of the Royal Society B: Biological Sciences,* 363, 717–739

Shennan, C., Cecchettini, C. L., Goldman, G. B. & Zalom, F. G. 2001. Profiles of California farmers by degree of IPM use as indicated by self-descriptions in a phone survey. *Agriculture, Ecosystems and Environment,* 84, 267–275

Sherwood, S., Cole, D., Crissman, C. & Paredes, M. 2005. Transforming potato systems in the andes. In: *The pesticide detox,* Pretty, J. (ed.), pp. Earthscan, London

Simpson, B. M. & Owens, M. Year. Farmer field schools and the future of agricultural extension in Africa. In: Proceedings of the 18th annual conference on approaches and partnerships for sustainable extension and rural development, 2002 Durban, South Africa, May 26 - 30, 2002. 405-412.

StockholmConvention. 2010. *UN chemical body recommends elimination of the toxic pesticide endosulfan*, Stockholm Convention on persistent organic pollutants (POPs), 05.01.2011, Available from:
http://chm.pops.int/Convention/Media/Pressreleases/POPRC6Geneva19Oct201
0/tabid/1042/language/en-US/Default.aspx

Thiam, M. & Touni, E. 2009. Pesticide poisoning in West Africa. *Pesticides News, 85*, 3-4

Thies, C. & Tscharntke, T. 1999. Landscape structure and biological control in agroecosystems. *Science, 285,* 5429, pp. 893–895

Tscharntke, T., Bommarco, R., Clough, Y., Crist, T. O., Kleijn, D., Rand, T. A., Tylianakis, J. M., Nouhuys, S. v. & Vidal, S. 2008. Conservation biological control and enemy diversity on a landscape scale. *Biological Control, 45,* 2, pp. 238–253

UNEP & UNCTAD. 2008. *Organic agriculture and food security in Afrca.* UNEP-UNCTAD Capacity Building tast Force on Trade, Environment and Development (CBTF). United Nations Environment Program and United Nations Conference on Trade and Development New York & Geneva

Uphoff, N. 1998. Understanding social capital: learning from the analysis and experience of participation. In: *Social Capital: A Multiperspective Approach*, Dasgupta, P. & Serageldin, I. (eds.), pp. World Bank, Washington DC

Waage, J. 1996. Yes, but does it work in the field? The challenge of technology transfer in biological control. *Entomophaga, 41,* 315-332

Wardle, D. A. 2006. The influence of biotic interactions on soil biodiversity. *Ecology Letters, 9,* 7, pp. 870–886

WB. 2004. *Volume 3. Demand- Driven Approaches to Agriculture Extension Case Studies of International Initiatives.* The World Bank, Washington

WB. 2006. *Toxic Pollution from Agriculture - An Emerging Story. Research in Vietnam and Bangladesh sheds new light on health impacts of pesticides*, The World Bank, 28.12.2010, Available from: http://go.worldbank.org/KN1TB1MO10

WHO. 1990. *Public Health Impact of Pesticides Used in Agriculture.* World Health Organization,

Williamson, S. 1998. Understanding natural enemies; a review of training and information in the practical use of biological control. *Biocontrol News and Information, 19,* 4, pp. 117-126

Williamson, S. 2003. Economic costs of pesticide reliance. *Pesticide news* 61, pp. 3-5

Williamson, S. F. J. 1997. *Farmer training capacity amongst Colombian institutions, with special reference to integrated management of coffee berry borer (broca). Final report on a consultancy mission to Colombia conducted between 2-24 March and 18-26 May 1997, on behalf of the DFID project: Integrated Pest Management for Coffee: Colombia, managed by the International Institute of Biological Control,* IIBC, Ascot, UK

Zossou, E. E. B. E. 2004. *Analyse des déterminants socio-économiques des practiques phytosanitaries: Cas des cultures maraîcheres à Cotonou.* Diplôme d'ingénieur agronome Diploma engineer of agronomy, University Abomey-Calavi.

3

Biodegradation of Pesticides

André Luiz Meleiro Porto, Gliseida Zelayarán Melgar,
Mariana Consiglio Kasemodel and Marcia Nitschke
Universidade de São Paulo, Instituto de Química de São Carlos,
Brazil

1. Introduction

The rapidly growing industrialization along with an increasing population has resulted in the accumulation of a wide variety of chemicals. Thus, the frequency and widespread use of man-made "xenobiotic" chemicals has led to a remarkable effort to implement new technologies to reduce or eliminate these contaminants from the environment. Commonly-used pollution treatment methods (e.g. land-filling, recycling, pyrolysis and incineration) for the remediation of contaminated sites have also had adverse effects on the environment, which can lead to the formation of toxic intermediates (Debarati et al., 2005). Furthermore, these methods are more expensive and sometimes difficult to execute, especially in extensive agricultural areas, as for instance pesticides (Jain et al., 2005). One promising treatment method is to exploit the ability of microorganisms to remove pollutants from contaminated sites, an alternative treatment strategy that is effective, minimally hazardous, economical, versatile and environment-friendly, is the process known as bioremediation (Finley et al., 2010).

Thereafter, it was discovered that microbes have the ability to transform and/or degrade xenobiotics, scientists have been exploring the microbial diversity, particularly of contaminated areas in search of organisms that can degrade a wide range of pollutants.

Hence, biotransformation of organic contaminants in the natural environment has been extensively studied to understand microbial ecology, physiology and evolution due to their bioremediation potential (Mishra et al., 2001). The biochemical and genetic basis of microbial degradation has received considerable attention. Several genes/enzymes, which provide microorganisms with the ability to degrade organopesticides, have been identified and characterized.

Thus, microorganisms provide a potential wealth in biodegradation. The ability of these organisms to reduce the concentration of xenobiotics is directly linked to their long-term adaptation to environments where these compounds exist. Moreover, genetic engineering may be used to enhance the performance of such microorganisms that have the preferred properties, essential for biodegradation (Schroll et al., 2004).

About 30% of agricultural produce is lost due to pests. Hence, the use of pesticides has become indispensable in agriculture.

The abusive use of pesticides for pest control has been widely used in agriculture. However, the indiscriminate use of pesticides has inflicted serious harm and problems to humans as well as to the biodiversity (Gavrilescu, 2005; Hussain et al., 2009). The problem of environmental contamination by pesticides goes beyond the locality where it is used. The agricultural pesticides that are exhaustively applied to the land surface travel long distances and can move downward until reaching the water table at detectable concentrations, reaching aquatic environments at significantly longer distances. Therefore, the fate of pesticides is often uncertain; they can contaminate other areas that are distant from where they were originally used. Thus, decontaminating pesticide-polluted areas is a very complex task (Gavrilescu, 2005).

Organochloride pesticides are synthetic and were widely used in the 1970s, mainly in the United States (http://www.epa.gov/history/topics/ddt/02.htm, accessed in May 2011). Although their use has been banished in many countries, they are still used in developing countries. Organochloride pesticides are cumulative in the organisms and pose chronic health effects, such as cancer and neurological and teratogenic effects (Vaccari et al, 2006). Many xenobiotic compounds are recalcitrant and resistant to biodegradation, especially the organochloride pesticides (Diaz, 2004; Dua et al., 2002; Chaudhry & Chapalamadugu, 1991). In general, these highly toxic and carcinogenic compounds persist in the environment for many years.

Organophosphorus pesticides are actually more widely used in the United States (http://www.chemicalbodyburden.org/cs_organophos.htm, accessed in May 2011). These pesticides affect the nervous system of insects and humans, in addition to influencing the reproductive system (Colosio et al., 2009; Jokanovic & Prostran, 2009). These chemical agents block the prolonged inhibition of the cholinesterase enzyme activity. These chemical agents block prolonged inhibition the activity of the enzyme cholinesterase (ChE), responsible for the nervous impulse in organisms (Yair et al., 2008). The excessive use of organophosphorus in agriculture has originated serious problems in the environment (Singh & Walker, 2006). Although, these pesticides degrade quickly in water, there is always the possibility that residues and byproducts will remain, in relatively harmful levels in the organisms (Silva et al., 1999; Ragnarsdottir, 2000).

Carbamate pesticides are important in the agriculture due to their broad activity spectrum. In addition to a wide range of compounds, they are relatively degraded and generally have a low degree of toxicity to humans (Wolfe et al., 1978). However, they inhibit the enzyme acetylcholinesterase, therefore they are considered toxic to humans. The inhibition of the hydrolysis reaction of acetylcholine (AcH) results in the accumulation of AcH, causing various symptoms, such as sweating, lacrimation, hypersalivation and convulsion of extremities (Suzuki & Watanabe, 2005).

Decontamination of pesticide-infested environments is a difficult matter and can be very costly. In fact, the damages from pesticides in the environment are practically irreparable. Any measure used to decrease the effects of pesticides in the environment will always be a palliative solution and never definitive for the problems caused. Regrettably, there is always irreparable damage to the organisms and the environment, as for instance, the extinction of bird species and microorganisms in the world.

The biological methods are advantageous to decontaminate areas that have been polluted by pesticides. These methods consider the thousands of microorganisms in the environment that in order to survive seek for alternatives to eliminate the pesticides that were sprayed. Many native microorganisms develop complex and effective metabolic pathways that permit the biodegradation of toxic substances that are released into the environment. Although the metabolic process is lengthy, it is a more viable alternative for removing the sources of xenobiotic compounds and pollution (Diaz, 2004; Schoefs et al., 2004; Finley et al., 2010).

On account of the grave risks synthetic pesticides pose to the organisms, there is an incessant search for pesticide safety and for the development of sustainable agriculture. The biological pesticides are based on natural compounds that effectively control the infestation of pests in agriculture. The advantage is that, contrary to synthetic pesticides, they are efficient and do not cause collateral damage (Fravel, 2005; Gerhardson, 2002; Raaijmakers et al., 2002).

The scope of this work demonstrates the use of the degradation of pesticides using microorganisms. This topic is inexhaustible and we are going to underscore the most recent points, including studies on the biodegradation of organochloride, organophophorus and carbamate pesticides by microbiological process. Afterwards, in perspective, this chapter will show the use of natural pesticides in the biological control of pests.

2. Biodegradation

According to the definition by the International Union of Pure and Applied Chemistry, the term biodegradation is "Breakdown of a substance catalyzed by enzymes *in vitro* or *in vivo*. This may be characterized for the purpose of hazard assessment such as:

1. Primary. Alteration of the chemical structure of a substance resulting in loss of a specific property of that substance.
2. Environmentally acceptable. Biodegradation to such an extent as to remove undesirable properties of the compound. This often corresponds to primary biodegradation but it depends on the circumstances under which the products are discharged into the environment.
3. Ultimate. Complete breakdown of a compound to either fully oxidized or reduced simple molecules (such as carbon dioxide/methane, nitrate/ammonium and water). It should be noted that the biodegradation products can be more harmful than the substance degraded." (http://sis.nlm.nih.gov/enviro/glossaryb.html, accessed in May 2011; http://www.epa.gov/OCEPAterms/bterms.html, accessed in May 2011).

Microbial degradation of chemical compounds in the environment is an important route for the removal of these compounds. The biodegradation of these compounds, i.e., pesticides, is often complex and involves a series of biochemical reactions. Although many enzymes efficiently catalyze the biodegradation of pesticides, the full understanding of the biodegradation pathway often requires new investigations. Several pesticide biodegradation studies have shown only the total of degraded pesticide, but have not investigated in depth the new biotransformed products and their fate in the environment.

2.1 Organochorine pesticides
2.1.1 Introduction

The organochlorine pesticides are known to be highly persistant in the environment. This class of pesticides includes the chlorinated derivatives of diphenyl ethane (dichlorodiphenyltrichloroethane - DDT, its metabolites dichlorodiphenyldichloroethylene - DDE, dichlorodiphenyldichloroethane - DDD, and methoxychlor), hexachlorobenzene (HCB), the group of hexachlorocyclohexane (α-HCH, β-HCH, Υ-HCH, δ-HCH, or lindane), the group of cyclodiene (aldrin, dieldrin, endrin, chlordane, nonachlor, heptachlor and heptachlor-epoxide), and chlorinated hydrocarbons (dodecachlorine, toxaphene, and chlordecone), (Menone et al., 2001; Patnaik, 2003). Figure 1 shows some structures of organochlorine pesticides.

Unlike the organophosphate and the carbamate pesticides, the toxic properties of the organochlorine pesticides are not very similar (Matolcsy et al., 1988). Although the toxicological properties are analogous to organochlorines with similar structures, like heptachlor and chlordane, the toxicological degree can vary by substituting a chlorine in the molecule. For instance, the substitution of chlorine atoms in the DDT ring for a methoxide group decreases the toxicity (Patnaik, 2003).

DDT is the most well known pesticide from the organochlorine group. The use of organochlorine pesticides started in 1939, when Paul Hermann Müller realized that the DDT, first synthesized by Othmar Zeidler in 1874, was an efficient insecticide (Matolcsy et al., 1988). The DDT's high efficiency, its low water solubility, its high persistence in the environment and its mode of action, unknown until that moment, contributed to the increasing use of DDT (Konradsen et al., 2004).

The industrial manufacture of DDT is based on the synthesis described by Zeidler. Chloral, chloral alcoholate or chloral hydrate is reacted with chloro-benzene in the presence of sulfuric acid, oleum or chlorosulfonic acid. The products obtained from the synthesis reaction contain several impurities, including the *ortho-para* [1,1,1-trichloro-2-(*o*-chlorophenyl)-2-(*p*-chlorophenyl)ethane] and *ortho-ortho* [1,1,1-trichloro-2-(*o*-chlorophenyl)-2-(*o*-chlorophenyl)ethane] isomers of DDT, and 1,1-dichloro-2,2-bis-(*p*-chlorophenyl)ethane (*p,p'*-DDD), its *ortho-para* isomer, 1,1-dichloro-2-(*o*-chlorophenyl)-2-(*p*-chlorophenyl)ethane (*o,p'*-DDD). The 1,1,1-trichloro-2,2-bis-(*p*-chlorophenyl)ethane (*p,p'*-DDT), is about 70% of the product mixtures (Matolcsy et al., 1988).

The *p,p'*-DDT is resistant to light, atmospheric oxygen and weak inorganic acids, but is rapidly decomposed to the biologically inactive *p,p'*-DDE (Matolcsy et al., 1988; Ahrens & Weber, 2009).

During the World War II, powder DDT was pulverized on the population's skin to prevent epidemics of typhus transmitted by lice. The insecticide was also used in other countries to control the malaria-bearing mosquitoes (Konradsen et al., 2004). The use of DDT to control malaria bearing mosquitoes earned Müller the 1948 Nobel Prize in Medicine. After the war, the use of DDT was adopted as an agricultural pesticides (Benn & McAuliffe, 1975; Ottaway, 1982; Mariconi, 1985), the results were so impressive, that its use continued for 25 to 30 years in most countries. The problem occurred when DDT, like most organochlorine, reduced its efficiency, forcing the use of higher dosages. Consequently, large specialized laboratories sought to develop formulas which were characterized by greater efficiency and biodegradability (Turk, 1989).

At the end of the 1950s, the biologist Rachel Carson began to gather examples of environmental damages attributed to DDT (D'Amato et al., 2002). Between 1970 and 1980, DDT agricultural use was banned in most developed countries. In 2004, the Stockholm Convention, outlawed several persistent organic pollutants, such as aldrin, dieldrin, endrin, toxaphene, mirex and heptachlor, and restricted DDT use to vector control (Ahrens & Weber, 2009; Arisoy & Kolankaya, 1998). In 2009, lindane and chlordecone were added to the outlawed list by the Fourth Conference of the Parties (Ahrens & Weber, 2009).

Although most organochlorine were banned from some countries, organochorine pesticides are still widely studied due to their racalcitrant nature, that is, even after years since the use has been banned, organochlorine contaminated sites are not rare. Not to mention, that the DDT use is still allowed to control malaria bearing mosquitoes, even though, narrowly.

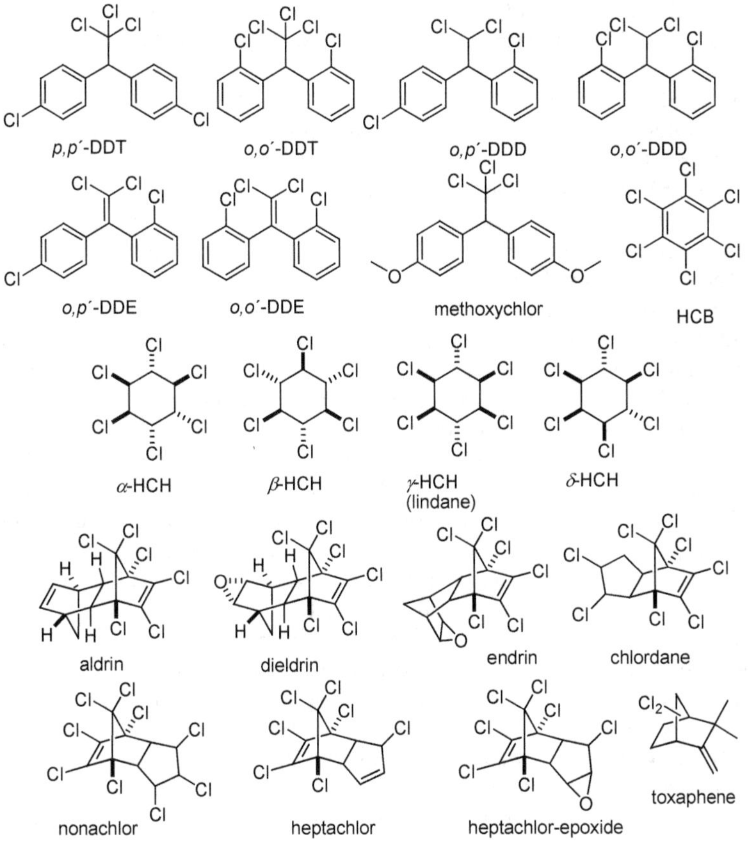

Fig. 1. Structures of organochlorine pesticides

2.1.2 Microbial degradation of organochloride pesticides

The fate of pesticides in the environment is determined by both biotic and abiotic factors. The rate at which different pesticides are biodegraded varies widely. Some pesticides such as DDT and dieldrin have proven to be recalcitrant. Consequently, they remain in the environment for a long time and accumulate into food chains for decades after their application to the soil (Kannan et al., 1994).

Most of the studies involving the biodegradation of organochlorine pesticides are done in pure cultures. The culture is usually isolated from a soil sample, generally contaminated with organochlorine pesticides. The strains are characterized and tested with different concentrations of the pesticide studied. DDT-metabolising microbes have been isolated from a range of habitats, including animal feces, soil, sewage, activated sludge, and marine and freshwater sediments (Johnsen, 1976; Lal & Saxena, 1982; Rochkind-Dubinsky et al., 1987).

The degradation of organochlorine pesticides by pure cultures has been proven to occur *in situ*. Nature magazine published one of the pioneer works. Matsumura et al. (1968) were able to evidence the breakdown of dieldrin in the soil by a *Pseudomonas* sp. The bacteria strain was isolated from a soil sample from the dieldrin factory yards of Shell Chemical

Company near Denver, Colorado. Later, in 1970, authors showed the biodegradation of aldrin, endrin and DDT with bacteria that were shown to be able to degrade dieldrin (Patil et al., 1970).

Biodegradation of DDT residues largely involves co-metabolism, that is, it requires the presence of an alternative carbon source, in which microorganisms growing at the expense of a substrate are able to transform DDT residues without deriving any nutrient or energy for growth from the process (Bollag & Liu, 1990).

Under reducing conditions, reductive dechlorination is the major mechanism for the microbial conversion of both the o,p'-DDT and p,p'-DDT isomers of DDT to DDD (Fries et al., 1969). The reaction involves the substitution of an aliphatic chlorine for a hydrogen atom. Using metabolic inhibitors together with changes in pH and temperature, Wedemeyer (1967) found that discrete enzymes were involved in the metabolism of DDT by *Aerobacter aerogenes*. The suggested pathway for the anaerobic transformation of DDT by bacteria is shown in Figure 2. Degradation proceeds by successive reductive dechlorination reactions of DDT to yield 2,2-bis(p-chlorophenyl)ethylene (DDNU), which is then oxidised to 2,2-bis(p-chlorophenyl)ethanol (DDOH). Further oxidation of DDOH yields bis(p-chlorophenyl)acetic acid (DDA) which is decarboxylated to bis(p-chlorophenyl)methane (DDM). DDM is metabolized to 4,4'dichlorobenzophenone (DBP) or, alternatively, may undergo cleavage of one of the aromatic rings to form p-chlorophenylacetic acid (PCPA). Under anaerobic conditions DBP was not further metabolized (Pfaender & Alexander, 1972). Through an investigation of the co-metabolism of DDT metabolites by a number of fungi (Subba-Rao & Alexander, 1985) were able to substantiate the pathway proposed by Wedemeyer (1967). There has been one report describing the conversion of DDE to 1-chloro-2,2-bis(p-chlorophenyl)ethylene - DDMU by bacteria (Masse et al., 1989).

Some studies have presented notable results on the biodegradation of organochlorine pesticides. Table 1 presents some of the microorganisms that were able to degrade organochlorine pesticides. Among microorganisms, bacteria comprise the major group involved in organochlorine degradation, especially soil habitants belonging to genera *Bacillus, Pseudomonas, Arthrobacter* and *Micrococcus* (Langlois et al., 1970). In order to predict some of the factors that influence the capacity of biodegradation of DDT by a *Sphingobacterium* sp., Fang et al. (2010), studied the biodegradation at different temperatures, pHs, concentrations of DDT and, with an additional source of carbon. Results of the experience showed that the degradation rates were proportional to the concentrations of p,p'-DDT, o,p'-DDT, p,p'-DDD and p,p'-DDE ranging from 1 to 50 mg.L^{-1}. The ability of *Sphingobacterium* sp. to degrade DDTs was somewhat inhibited by DDTs at the level as high as 50 mg.L^{-1}. According to the authors, this may be due to the fact that DDTs at high concentration are toxic to *Sphingobacterium* sp. and inhibit degradation. The experiment was also tested for different pHs, it was tested for pH 5, 7 and 9. The results indicated that a neutral condition is favorable for the degradation of DDT by *Sphingobacterium* sp., whereas higher or lower pH inhibits degradation. The influence of the temperature on the biodegradation was investigated by performing the experiments at temperatures of 20, 30 and 40 °C. The results indicated that the optimum temperature for the biodegradation of DDTs by a *Sphingobacterium* sp. in pure culture was at 30 °C. Ultimately, the biodegradation was available with an additional carbon source and results showed that the degradation half-lives of DDTs in the presence of glucose, yeast extract, sucrose, and frutose were significantly shorter than those in the treatment without an additional carbon source; and that the presence of glucose generates the fastest degradation of DDTs (Fang et al., 2010).

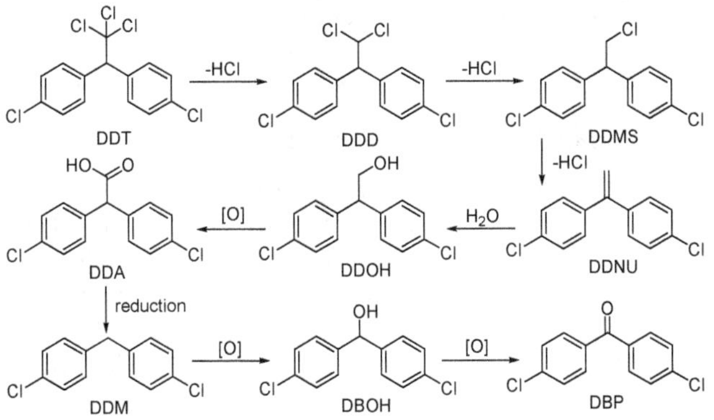

Fig. 2. Proposed pathway for bacterial metabolism of DDT (Adapted from Aislabie et al., 1997).

Results obtained *in vitro* can be applied to contaminated sites for further investigation on the capacity of a microorganism to degrade an organochlorine pesticide. As a continuation of the study proposed by Fang et al. (2010), the bacteria that was evidenced to degrade DDT was applied to field soils after different treatments. The soil known to be contaminated with DDT was studied in four different conditions, the control, which did not receive any treatment; PV, the same soil, only with pumpkin vegetation; DI received inoculation of the *Sphingobacterium* sp. and; PVDI which was the contaminated soil with the pumpkin vegetation and inoculation with *Sphingobacterium* sp. The concentration of p,p'-DDT, o,p'-DDT, p,p'-DDD and p,p'-DDE was measured from each soil sample after 90 days, and was then compared to the initial concentration. Analysis indicated that the removal percentages of o,p'-DDT and p,p'-DDE in the PVDI treatment were statistically significantly higher (Fang et al., 2010).

According to Aislabie & Jones (1995) and Aislabie et al. (1997), the microbial degradation of DDT in soil apparently proceeds by a pathway analogous to that proposed by Wedemeyer (1967), (Figure 2). Under anaerobic conditions the first and major biotransformation product of DDT is DDD, with minor levels of DDA, DDM, DDOH, DBP, and DDE being detected (Guenzi & Beard, 1967; Mitra & Raghu, 1988; Xu et al.,1994; Boul et al. ,1994). Reports of biodegradation of DDE in soil are rare, although, Agarwal et al. (1994) described the isolation of DDMU as a biotransformed product of DDE.

Studies with fungi have also evidenced the biodegradation of organochlorine pesticides. Ortega et al. (2011) evaluated marine fungi collected off the coast of São Sebastião, North of São Paulo State, Brazil. The fungi strains were obtained from marine sponges. The fungi *Penicillium miczynskii*, *Aspergillus sydowii*, *Trichoderma* sp., *Penicillium raistrickii*, *Aspergillus sydowii* and *Bionectria* sp. were previously tested in solid culture medium containing 5, 10 and 15 mg of DDD. The tests were also carried out with liquid medium in a rotary shaker, with the same amount of DDD per 100 mL liquid medium. The results showed that the fungi *P. miczynskii, A. sydowii* and *Trichoderma* sp. presented good growth in the presence of the pesticide. For further experiments *Trichoderma* sp. was selected as the standard microorganism, as it showed the best resistance to DDD in both solid and liquid medium (Ortega et al., 2011). In the experiments where DDD pesticide was concomitantly added into the growth of *Trichoderma* sp., 21% of the pesticide was degraded. The addition of H_2O_2 in the experiment promoted a degradation increase (75%). In the experiments where DDD was added after 5 days of *Trichoderma* sp. growth, and with the addition of H_2O_2, the total biodegradation occurred (Ortega et al., 2011). Many factors can affect the biodegradation, as described earlier, as for instance the presence of H_2O_2 increases the efficiency of the DDD degradation by *Trichoderma* sp.

Pesticides	Toxicity[1]	Microorganisms	References
PCP	Class Ib	*Arthrobacter* sp.	(Stanlake & Finn , 1982)
		Flavobacterium sp.	(Crawford & Mohn, 1985)
1,4-Dichlorobenzene	Class II	*Pseudomonas* sp.	(Spain & Nishino, 1987)
DDT	Class II	*Aerobacter aerogenes*	(Wedemeyer, 1966)
		Trichoderma viridae	(Patil et al., 1970)
		Pseudomonas sp.	(Patil et al., 1970)
		Micrococcus sp.	(Patil et al., 1970)
		Arthrobacter sp.	(Patil et al., 1970)
		Bacillus sp.	(Patil et al., 1970)
		Pseudomonas sp.	(Kamanavalli & Ninnekar, 2005)
		Sphingobacterium sp.	(Fang et al., 2010)
Lindane	Class II	*Basea thiooxidans*	(Pesce & Wunderlin, 2004)
		Sphingomonas paucimobilis	(Pesce & Wunderlin, 2004)
		Streptomyces sp.	(Benimeli et al., 2008)
		Pleurotus ostreatus	(Rigas et al., 2005)
DDE	n.i.	*Phanerochaete chrysosporium*	(Bumpus et al., 1993)
DDD	n.i.	*Trichoderma* sp.	(Ortega et al., 2011)
Heptachlor epoxide	n.i.	*Phlebia* sp.	(Xiao et al., 2011)
Heptachlor	O	*Phanerochaete chrysosporium*	(Arisoy & Kolankaya, 1998)
		Phlebia sp.	(Xiao et al., 2011)
Toxaphene	O	*Bjerkandera* sp.	(Lacayo et al., 2006)
Aldrin	O	*Trichoderma viridae*	(Patil et al., 1970)
		Pseudomonas sp.	(Patil et al., 1970)
		Micrococcus sp.	(Patil et al., 1970)
		Bacillus sp.	(Patil et al., 1970)
Endrin	O	*Trichoderma viridae*	(Patil et al., 1970)
		Pseudomonas sp.	(Patil et al., 1970)
		Micrococcus sp.	(Patil et al., 1970)
		Arthrobacter sp.	(Patil et al., 1970)
		Bacillus sp.	(Patil et al., 1970)
Dieldrin	O	*Pseudomonas* sp.	(Matsumura et al., 1968)

Table 1. Microorganisms involved in degradation of organochlorine pesticides.

[1]WHO Recommended classification of pesticides by Hazard. Class I is subdivided into two other classifications: Class Ia (extremely hazardous) and Class Ib (highly hazardous), Class II are the moderately hazardous, Class III are slightly hazardous and Class u, are unlikely to present acute hazard. The use of some pesticides have been discontinued and are classified as obsolete (O), (WHO, 2009).

n.i. – not informed

Studies involving the biodegradation of polychlorinated biphenyls (PCBs), used as a pesticide extender, have also been conducted, several isolated microorganisms have been proven to be capable of aerobically degrade PCBs, preferentially degrading the more lightly chlorinated congeners. These organisms attack PCBs via 2,3-dioxygenase pathway, converting PCBs to the corresponding chlorobenzoic acids. These chlorobenzoic acids can then be degraded by indigenous bacteria, resulting in the production of carbon dioxide, water, chloride, and biomass (Abramowicz, 1995). Anaerobic bacteria attack more highly chlorinated PCB congeners through reductive dechlorination. In general, this microbial process removes preferentially the *meta* and *para* chlorines, resulting in a depletion of highly chlorinated PCB congeners with corresponding increases in lower chlorinated, *ortho*-substituted PCB congeners (Abramowicz, 1995).

Despite the evidence that microorganisms with the ability to degrade DDT are resident in soil, its residues persist. The studies here presented showed that anaerobic conditions are beneficial to dechlorination of DDT, and additional carbon and hydrogen peroxide favors the biodegradation of some organochlorines. The decomposition rate depends on conditions in the soil and the bonding of the pesticide to soil surfaces. For most pesticides, aerobic decomposition proceeds much faster than anaerobic decomposition; however, there are classic exceptions to this, for instance DDT, whose decomposition proceeds ten times faster under anaerobic conditions (Scott, 2000).

Farm management practices also affect the rate at which pesticides are degraded. Irrigation of soils has been shown to enhance degradation of DDT to DDD, thought to be due to the creation of anaerobic microsites (Aislabie & Jones, 1995).

Due to the recalcitrant nature of most organochlorines, many such pesticides are still widely studied in order to find mechanisms that enhance their biodegradation in the environment. Genetic techniques can contribute to elucidate biochemical pathways involved in the microbial degradation of organochlorines, which represent promising alternatives towards developing highly efficient strains as well as the isolation and application of enzymes potentially involved in biodegradation.

2.2 Organophosphate pesticides
2.2.1 Introduction

Currently, among the various groups of pesticides that are used worldwide, organophosphorus pesticides form the major and most widely used group that accounts for more than 36% of the total world market. The most used among these is methyl parathion. Its accumulation has many health hazards associated to it, hence, its degradation is very important (Ghosh et al., 2010).

The organophosphorus pesticides (OP) are all esters of phosphoric acid and are also called organophosphates, which include aliphatic, phenyl and heterocyclic derivatives (Figure 3). Owing to large-scale use of OP compounds, contaminations of soil and water systems have been reported from all parts of the world. In light of this, bioremediation provides a suitable way to remove contaminants from the environment as, in most cases, OP compounds are totally mineralized by the microorganisms. Most OP compounds are degraded by microorganisms in the environment as a source of phosphorus and /or carbon. Classification of Pesticides. Thus, the OP pesticides can be hydrolyzed and detoxified by carboxylesterase and phosphotriesterase enzymes.

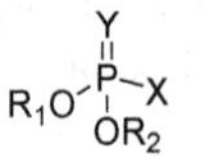

R₁, R₂ = Methyl, Ethyl

Y = O, S

X = Specific Organic Group

Fig. 3. General structure of organophosphate pesticides

Organophosphates are used to control a variety of sucking, chewing and boring insects, spider mites, aphids, and pests that attack crops like cotton, sugarcane, peanuts, tobacco, vegetables, fruits and ornamentals. OP pesticides are marketed by many of the world's major agrochemical companies. Some of the main agricultural products are parathion, methyl parathion, chlorpyriphos, malathion, monochrotophos, diazinon, fenitrothion and dimethoate (Figure 4).

The organophosphorates possess an efficient insecticide activity, due to its characteristic of irreversibly inhibiting the enzyme acetylcholinesterase in the nervous system, which acts in both insects and in mammal. In man, the organophosphates are absorbed through all routes, reaching high concentrations in fatty tissues, liver, kidneys, salivary glands, thyroid, pancreas, lungs, stomach, intestines and, at smaller proportions, in the central nervous system (SNC) and muscles. However, the organophosphates do not accumulate in the human organism, as it is readily biotransformed in the liver. The excretion of these compounds and of their metabolites is quite fast, taking place mostly in the urine and, at small proportions, in the feces, usually within 48 h. The largest excretion levels occur within 24 h after absorption (Oga, 2003; Griza et al., 2008).

Due to the above mentioned health hazards and other problems associated with the use organophosphorus pesticides, early detection and subsequent decontamination and detoxification of the polluted environment is essential. The present subject examines applications and future use of OP-degrading microorganism cultures from agricultural fields and enzymes for bioremediation (Karpouzas & Singh, 2006).

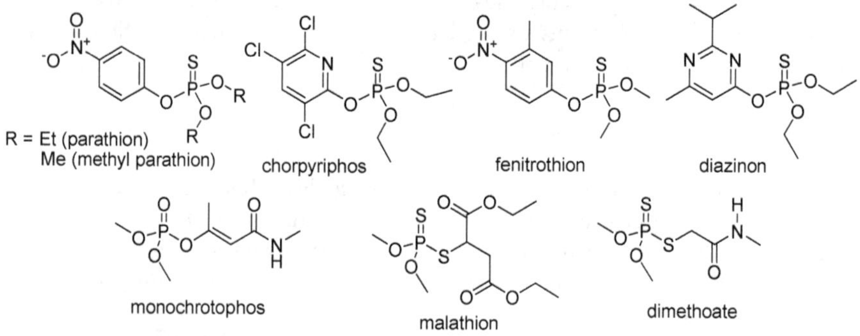

Fig. 4. Structures of organophosphorate pesticides

2.2.2 Microbial degradation of organophosphate pesticides

Methyl parathion (O,O-dimethyl-O-(p-nitro-phenylphosphorothioate) is one of the most used organophosphorus pesticides. This product is widely used throughout the world and its residues are regularly detected in a range of fruits and vegetables. Investigation of microbial degradation is useful for developing insecticide degradation strategies using microorganisms. Bacteria with the ability to degrade methyl parathion have been isolated worldwide (Liu et al., 2003; Hong et al., 2005).

Multiplex tendencies characterize pesticide applications in farming. A number of pesticide mixtures, especially pyrethroid and organophosphorus pesticide mixtures have been formulated as an improvement over individual pesticides (Moreby et al., 2001). Construction of a genetically engineered microorganism (GEM), which can simultaneously degrade these two kinds of pesticides, could benefit the study and application of bioremediation in multiple pesticide-contaminated environments (Yuanfan et al., 2010). Methyl parathion hydrolase gene, *mpd*, which is responsible for hydrolyzing methyl parathion to *p*-nitrophenol and dimethyl phosphorothioate, has also been cloned from these strains. Sequences are effectively conserved in these strains (Yuanfan et al., 2010).

A fenpropathrin-degrading bacterium, *Sphingobium* sp. JQL4-5, was isolated and characterized. A stable, genetically engineered strain, JQL4-5-*mpd*, capable of simultaneously degrading fenpropathrin and methyl parathion was constructed by random insertion of the methyl parathion hydrolase gene (*mpd*) into the chromosome of strain JQL4-5. Soil treatment results indicated that JQL4-5-*mpd* is a promising GEM in the bioremediation of multiple pesticide contaminated environments (Yuanfan et al., 2010).

Organophosphorus hydrolase (OPH), isolated from both *Flavobacterium* sp. ATCC 27551 (Mulbry & Karns, 1989) and *Pseudomonas diminuta* MG (Serdar et al., 1989), is capable of hydrolyzing a wide range of oxon and thion OPs. However, OPH has already been shown to lack any hydrolytic activity toward numerous dimethyl OPs (Horne et al., 2002). The *mpd* gene encoding an organophosphate degrading protein was isolated from a methyl parathion (MP) degrading *Plesiomonas* sp.

The methyl parathion hydrolase gene (*mpd*) and enhanced green fluorescent protein gene (*egfp*) was successfully coexpressed using pETDuet vector in *Escherichia coli* BL21 (DE3). The coexpression of methyl parathion hydrolase (MPH) and enhanced green fluorescent protein (EGFP) were confirmed by determining MPH activity and fluorescence intensity. The recombinant protein MPH showed high enzymatic degradative activity of several widely used OP residues on vegetables. Subsequently, a dual-species consortium comprising engineered *E. coli* and a natural *p*-nitrophenol (PNP) degrader *Ochrobactrum* sp. strain LL-1 for complete mineralization of dimethyl OPs was studied. The dual-species consortium possesses the enormous potential to be utilized for complete mineralization of PNP-substituted OPs in a laboratory-scale bioreactor. These studies demonstrated that MP could be degraded via the MP → PNP → hydroquinone → Krebs cycle (Figure 5) by the dual-species consortium. The data confirm that the mineralization process of MP is initiated by hydrolysis leading to the generation of PNP and dimethylthiophosphoric acid, and PNP degradation, then, proceeds through the formation of hydroquinone. The accumulation of PNP in suspended culture was prevented (Zhang et al., 2008).

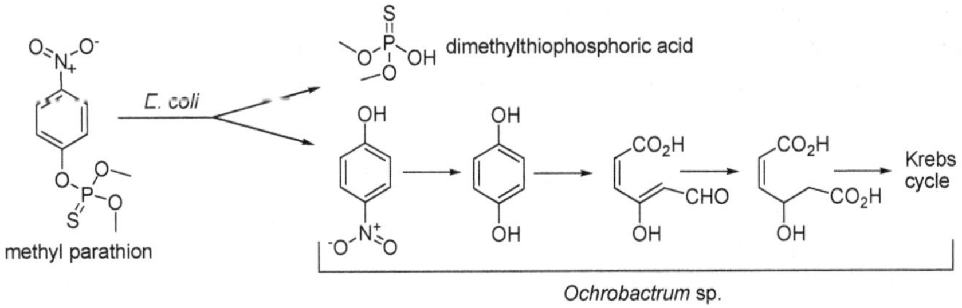

Fig. 5. Proposed pathway for the biodegradation of MP by microbial consortium (Zhang et al., 2008)

Thus, there is an increasing need to develop new methods to detect, isolate, and characterize the strains/enzymes playing a part in these degradation processes (Vallaeys et al., 1996). Successful detoxification of recalcitrant organic chemicals may require the concerted effort of multispecies consortia.

Fenamiphos (FEN), ethyl 4-methylthio-*m*-tolyl isopropylphosphoramidate, is an organophosphate nematicide used in protected horticultural crops. In soil, FEN is gradually oxidized to its sulfoxide (FSO) and sulfone (FSO$_2$), which also possess high nematicidal activity and is equally toxic to non-target vertebrates (Figure 6). Degradation studies of FEN in a range of soils showed half-life values ranging from 12 to 87 days. FEN and its oxidation products FSO and FSO$_2$ showed low to moderate affinity for soil adsorption and their soil accumulation may result in their eventual downward movement into groundwater. Indeed, previous studies have suggested that under favorable environmental conditions FEN could leach to groundwater where it could persist (Franzmann et al., 2000). Therefore, tools are needed for the decontamination of natural resources by the residues of chemicals such as FEN and its oxidation derivatives.

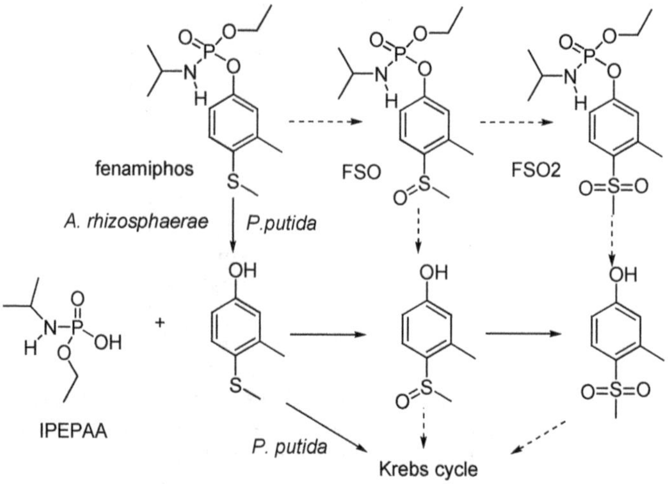

Fig. 6. Metabolic pathway of FEN by the isolated bacteria (Chanika et al., 2011)

Two bacteria identified as *Pseudomonas putida* and *Acinetobacter rhizosphaerae*, able to rapidly degrade the organophosphate fenamiphos, were isolated. Denaturing gradient gel electrophoresis analysis revealed that the two isolates were dominant members of the enrichment culture. Clone libraries further showed that bacteria belonging to α-, β-, γ-Proteobacteria and Bacteroidetes were also present in the final enrichment, but were not isolated. Both strains hydrolyzed FEN to fenamiphos phenol and ethyl hydrogen isopropylphosphoramidate (IPEPAA), which was further transformed, only by *P. putida*. The two strains were using FEN as C and N source. Cross-feeding studies with other pesticides showed that *P. putida* degraded OPs with a P–O–C linkage (Chanika et al., 2011).

Thus, both bacteria were able to hydrolyze FEN, without prior formation of FSO or FSO2, to FEN-OH which was further transformed only by *P. putida* (Figure 6), suggesting elimination of environmentally relevant metabolites. In addition, *P. putida* was the first wild-type bacterial isolate able to degrade OPs. All the above characteristics of *P. putida* and its demonstrated ability to remove aged residues of FEN highlight its high bioremediation potential (Chanika et al., 2011).

Herein, it was shown that the construction of genetically engineered microorganism (GEM) and the dual-species consortium has the potential to be used in the degradations of different kinds of pesticides. These studies show the benefits of bioremediation in multiple pesticide-contaminated environments and mineralization of toxic intermediates in the environment, which can lead to complete bioremediation of contaminated sites that have an adverse effect.

2.3 Carbamate pesticides
2.3.1 Introduction

Carbamates were introduced as pesticides in the early 1950s and are still used extensively in pest control due to their effectiveness and broad spectrum of biological activity (insecticides, fungicides, herbicides). High polarity and solubility in water and thermal instability are typical characteristics of carbamate pesticides, as well as high acute toxicity. The carbamates are transformed into various products in consequence of several processes such as hydrolysis, biodegradation, oxidation, photolysis, biotransformation and metabolic reactions in living organisms (Soriano et al., 2001).

Chemically, the carbamate pesticides are esters of carbamates and organic compounds derived from carbamic acid (Figure 7). This group of pesticides can be divided into benzimidazole-, N-methyl-, N-phenyl-, and thiocarbamates. The compounds derived from carbamic acid are probably the insecticides with the widest range of biocide activities (Sogorb & Vilanova, 2002).

carbamic acid   carbamates

Fig. 7. General structures of carbamate pesticides

Highly toxic acetylcholinesterase (AChE)-inhibiting pesticides, organophosphates and carbamates are intensively used throughout the world and continue to be responsible for poisoning epidemics in various countries (De Bleecker, 2008). The carbamates are inhibitors of AChE and are responsible for the greatest number of poisonings in the rural environment. The use of pesticides in the Brazilian rural environment has brought a series of dire consequences to the environment as well as to the health of rural workers (Oliveira-Silva et al., 2001). The clinical effects of carbamate pesticides depend on the dose, route of exposure, type of carbamate involved, use of protective gear, and the premorbid state of the victim (Rosman et al., 2009).

The study of pesticide degradation is usually beneficial, since the reactions that destroy pesticides convert most pesticide residues in the environment to inactive, less toxic, harmless compounds (Lan et al., 2006).

The enzymatic hydrolysis of carboxyl esters by carboxyl esterases (CbEs) is based on the reversible acylation of a serine residue within the active centre of the protein (Gupta, 2006). Firstly, the substrate must gain access to the active site and this acylation causes a nucleophilic attack by the serine on the carboxyl carbamate producing a transition state formation, in addition to forming a stable acylated enzyme (Figure 8). This acyl-enzyme intermediate is hydrolysed by nucleophilic attack of water that releases the corresponding carbamine acid, plus the free active enzyme again ready to initiate a new catalytic cycle (Reed & Fukuto, 1973; Sogorb & Vilanova, 2002; Hemmert & Redinbo, 2010).

Moreover, the investigation of biodegradation pathways are quite complex. In addition to the complexity of the (bio)degradation of the pesticides there are also other factors, such as pesticide nonextractable residues in soils. The definition of bound residues was described as "bound residues represent compounds in soils, plants, or animals which persist in the matrix in the form of the parent substance or its metabolite(s) after extraction. The extraction method must not substantially change the compounds themselves or the structure of the matrix" (Barriuso et al., 2008).

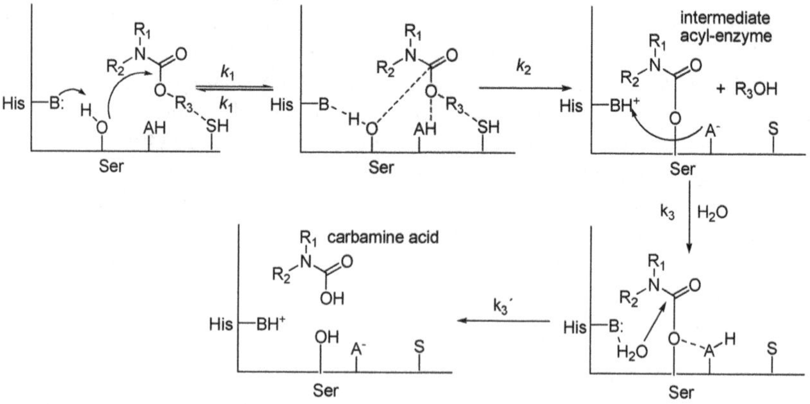

Fig. 8. Catalytic mechanism for carbamate hydrolysis by carboxyl esterases, k_1 defines the affinity of the enzyme for a given substrate and k_2 describes how quickly the acyl-enzyme intermediate is formed (Reed & Fukuto, 1973).

2.3.2 Microbial degradation of carbamate pesticides

The biodegradation of carbamates has been investigated by different microorganisms that metabolize carbamate pesticides. In most cases, the studies did not eliminate the possibility that abiotic processes are involved in the degradation.

A number of bacteria capable of degrading carbofuran (*Pseudomonas*, *Flavobacterium*, *Achromobacterium*, *Sphingomonas*, *Arthrobacter*) have been isolated and characterized in an effort to better understand the bacterial role to remove carbofuran from the environment. Carbofuran is one of the pesticides belonging to the N-methylcarbamate class used extensively in agriculture. It exhibits high mammalian toxicity and has been classified as highly hazardous. Carbofuran was degraded first to carbofuran phenol and the result was degraded to 2-hydroxy-3-(3-methylpropan-2-ol) phenol by *Sphingomonas* sp. (Kim et al., 2004) and *Arthrobacter* sp. (De Schrijver & De Mot, 1999), (Figure 9).

Carbendazim is a widely used broad-spectrum benzimidazole fungicide to control a wide range of fungal pathogens on cereals and fruits, it is also used in soil treatment and foliar application on the appearance of disease. The fungicide carbendazim was degraded by a microbial consortium obtained from several soil samples in Japanese paddy fields with continuous culture enrichment. Biodegradation using immobilized bacterial consortium was investigated in various parameters, as temperature, pH, and nutrient concentration. The degradation ability of the consortium was increased by immobilization on loofa (*Luffa cylindrica*) sponge, in comparison with that of free-living consortium. This immobilized consortium on loofa sponge is a promising material for bioremediation of polluted water with these pesticides in paddy fields (Pattanasupong et al., 2004).

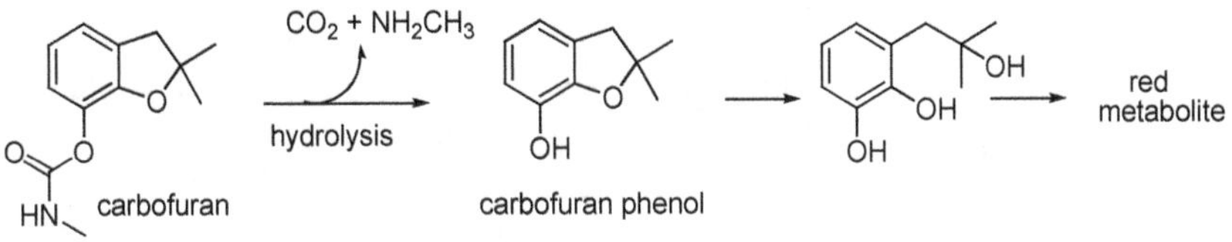

Fig. 9. Biodegradation of carbofuran by *Sphingomonas* sp.

Afterwards, the carbamate carbendazin was converted to 2-aminobenzimidazole by *Pseudomoans* isolates (Figure 10). In general, a limited number of xenobiotic pesticides are metabolized by single strain, but usually consortia of microorganisms are catalyzed for complete degradation. Several Actinomycetes that metabolize carbamate pesticides were isolated. In most cases, this is initiated by hydrolysis of the carbamate at the ester linkage. (De Schrijver & De Mot, 1999).

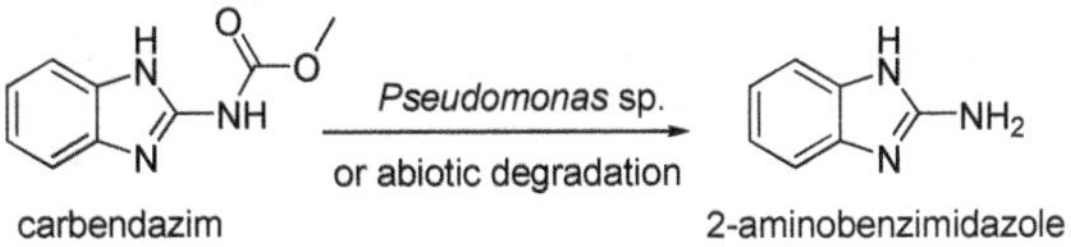

Fig. 10. Biodegradation of carbendazim by *Pseudomonas* sp.

Juvenoids are efficient pesticides with relatively low toxicity to humans (Figure 11). However, few studies have evaluated the effect of degradation by soil microorganisms on their toxicity. The effects of bacterial, fungal and yeast isolates on aerobic decomposition of ethyl N-{2-[4-(2,2-ethylenedioxy-1-cyclohexylmethyl)phenoxy]ethyl} carbamate during eight weeks were determined. Higher degradation activity was observed during the first week of the experiment and a substantial decrease in the rate of degradation occurred during the following seven weeks. This can be described both to the accumulation of degradation products and to impaired physiological state of microbial cultures during the long-term experiments (Novák et al., 2003).

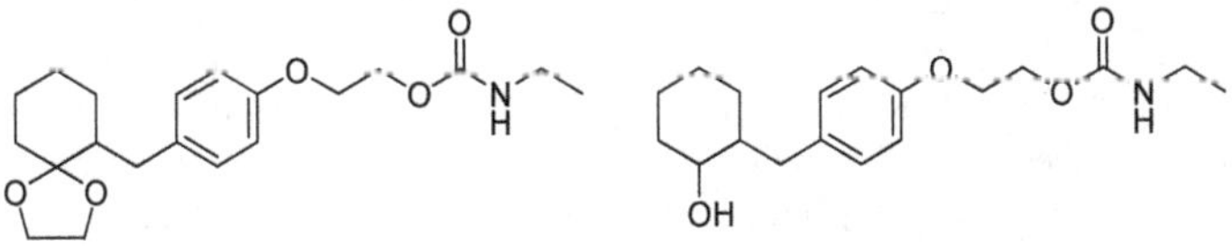

Fig. 11. Juvenoid pesticides

Ethylenethiourea is an important degradation product of ethylenebisdithiocarbamate fungicides (maneb, zineb, mancozeb), which are widely used in different kinds of crops (Figure 12). The ethylenebisdithiocarbamates are not highly toxic and degrade rapidly in the presence of moisture and oxygen, forming different types of compounds such as the polar ethylenethiourea, which is relatively stable and is a potential contaminant for groundwater. Experiments conducted under biotic and abiotic conditions, showed complete degradation of ethylenethiourea in the presence of microbial nitrate reduction with pyrite, which occurs in deeper parts of the aquifers (Jacobsen & Bossi, 1997).

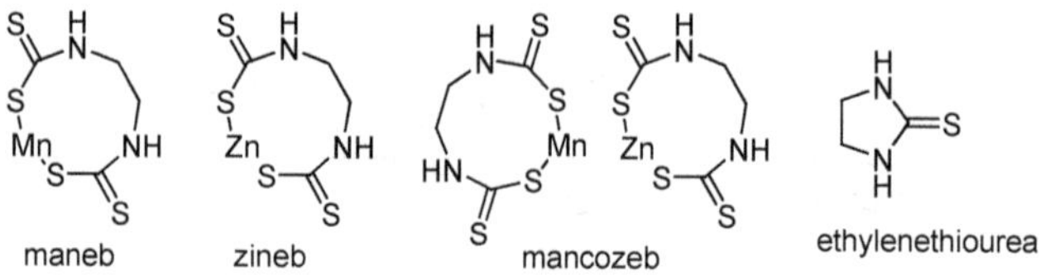

maneb   zineb   mancozeb   ethylenethiourea

Fig. 12. Ethylenethiourea pesticides

In general, pesticide-degrading microorganisms are isolated via enrichment cultures. A novel strategy has been reported using a coexpression vector for the purpose of developing bacteria that can detoxify different pesticides. The organophosphate hydrolase gene from *Flavobacterium* sp. and carboxylesterase B1 gene (b1) from *Culex pipiens* were cloned in the coexpression vector. A single microorganism was capable of producing both enzymes for degradation of organophosphate (parathion), carbamate (pirimicarb) and pyrethroid pesticides (deltamethrim), (Figure 13). The technical capability of genetically engineering bacteria with more enzymes should open up new opportunities for extending the wide range of pesticides that can be biodegraded in the future (Lan et al., 2006).

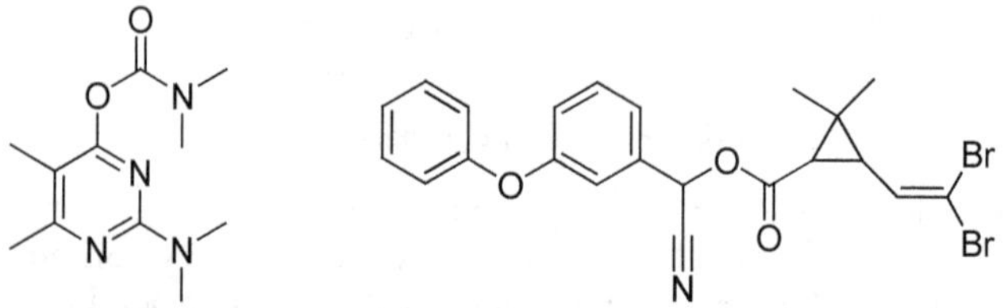

Fig. 13. Pirimicarb (left) and deltamethrim (right) pesticides

Recently, the isolation of a soil bacteria able to hydrolyze organophosphate and carbamate pesticides was performed. Cross-feeding studies with other pesticides showed that *Pseudomonas putida* degraded organophosphates with a P–O–C linkage (fenamiphos), and oxamyl (Figures 6 and 14) and carbofuran carbamates (Chanika et al., 2011). In addition, the biodegradation of insecticidal organophosphates and carbamates has been described by human brain esterases, which actively degraded 1-naphthyl acetate and other substrates (Sakai & Matsumura, 1971).

Contamination of surface water by organophosphate and carbamate compounds is of concern because of the potential toxicity to aquatic organisms, especially those at lower trophic levels. Many organophosphate and carbamate compounds have acute and chronic toxicity to fish and aquatic invertebrates. Bondarenko et al. (2004) showed that the persistence of diazinon and chlorpyrifos was much longer than for malathion and carbaryl in freshwater, and was further prolonged in seawater. Afterwards, microbial degradation contributed significantly to the dissipation of diazinon and chlorpyrifos in freshwater, but was inhibited in seawater. In contrast, degradation of malathion and carbaryl was rapid and primarily abiotic. The interactions of pesticide persistence with water location, temperature, and type of pesticides suggest that site, and compound-specific, information is needed when evaluating the overall ecotoxicological risks of pesticide pollution in a watershed.

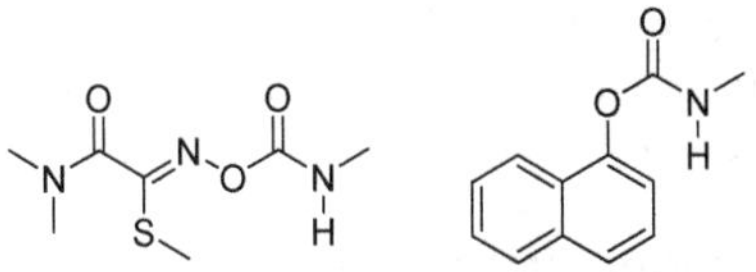

Fig. 14. Oxamyl (left) and carbaryl (right) pesticides

Kaufman and Blake (1973) have selected soil microorganisms capable of degrading isopropyl carbanilate (propham), 3',4'-dichloropropionanilide (propanil), 3'-chloro-2-methyl-*p*-valerotoluidide (solan), and methyl 3,4-dichlorocarbanilate (swep), (Figure 15). Degradation of the pesticides in enrichment solutions, and by pure cultures of effective microbial isolates (*Pseudomonas striata, Achromobacter* sp., *Aspergillus ustus, Aspergillus versicolor, Fusarium oxysporum, Fusarium solani, Penicillium chrysogenu, Penicillium janthinellu, Penicillium rugulosum* and *Trichoderma viride*) were demonstrated by the production of the corresponding aniline, chloride ion liberation and disappearance of the original compounds. Each organism demonstrated unique substrate specificity and was capable of degrading other aniline-based pesticides of the acetamide, acylanilide, carbamate, toluidine and urea classes.

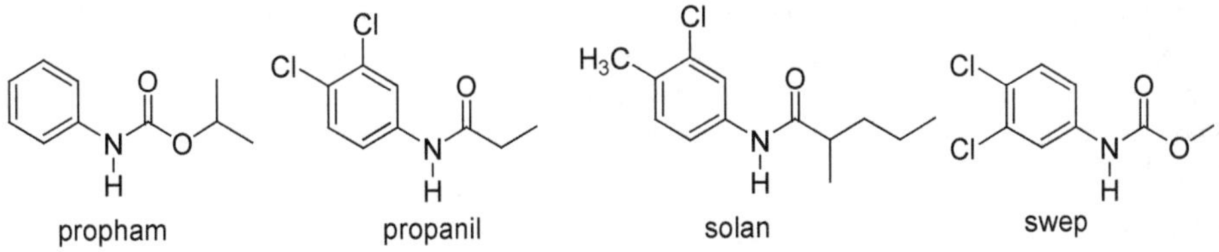

Fig. 15. Carbamate pesticides degraded by soil microorganisms

As described here, the carbamate pesticides are easily degraded by different types of microorganisms (fungi and bacteria). The degradation of these pesticides by enzymatic systems of microorganisms has contributed to the total removal of xenobiotics from soils, hence avoiding the contamination of waters and the environment.

2.4 Biological pesticides

Synthetic chemical pesticides provide many benefits to agriculture and food production, however, as previously discussed, they also present toxicity to non-target organisms and cause environmental pollution, therefore efforts to find new pest control alternatives have been studied, essentially due to the increasing concern about the effects of these compounds on human health and on the environment. Biodegradation and bioremediation of synthetic pesticides have been used as alternative green technologies to solve the problems related to the accumulation of these contaminants in soil and water. Another proposal to reduce the environmental impact of pesticides is the use of biological-derived products also known as biopesticides.

According to the Environmental Protection Agency (EPA), biopesticides are defined as naturally occurring pest control substances. They are classified into three groups (Joshi, 2006):

a. Microbial pesticides: in which a microbial living organisms (bacteria, fungi, viruses, protozoans) is the active control agent;
b. Plant pesticides: pesticidal substances produced by plants from introduced genetic material (plant incorporated protectants);
c. Biochemical pesticides: naturally occurring substances that control pests by non-toxic mechanisms. These include substances that interfere with growth or mating such as pheromones.

The main advantage of biopesticides is their safety to non-target organism, biodegradability and their specificity, which permits the use of small dosages and power exposure, hence avoiding pollution caused by conventional pesticides (Rosell et al., 2008). In addition to being less harmful than chemicals, biopesticides have been of great value in integrated pest management (IPM) strategies where the use of biopesticides greatly decreases the use of chemicals, maintaining crop yields. The specificity of biopesticides contrasts with the broad spectrum of chemical counterparts. In contrast, biopesticides are also slow acting, have relatively critical application times, most suppress rather than eliminate the target population, have limited field persistence and short-shelf life.

Despite the range of biopesticides that have been described, our discussion focuses on microbial pesticides.

2.4.1 Microbial pesticides

Microbiological control is sustained by beneficial interactions resulting from competition, antagonism and parasitism of microorganisms against plant pathogens, insects and weeds (Montesinos, 2003). In general, microorganisms are able to suppress pests by producing a toxin, causing a disease or preventing the establishment of other organisms. Currently, several microorganisms involved in such processes are the active ingredient of microbial pesticides.

2.4.1.1 Bacteria

Most biopesticides available in the market are bacterial-based products. The well-known and widely used bacterial biopesticide comprises the gram-positive, spore-forming bacteria belonging to the genus *Bacillus* that are commonly found in soil. The majority of commercial microbial insecticides are preparations based on strains of *Bacillus thuringiensis* (Bt) that produces a crystalline inclusion body during sporulation (Frankenhuyzen, 2009). The crystal proteins (Cry proteins) are toxic to many insects and are defined as endotoxins (Bt toxin) that are generally encoded by bacterial plasmids (Gonzales & Carlton, 1980). Both spores and inclusion bodies are released upon lysis of the parent bacterium at the end of the sporulation cycle and if ingested, the spores and crystals act as poisons in certain insects. The protein is activated by alkaline conditions and enzyme activity of insect's gut hence, Bt is referred as a stomach poison (Chattopadhyay et al., 2004). The toxicity of the activate protein is dependent on the presence of receptor sites on the insects gut wall. This match between toxin and receptor sites determines the range of insect species killed by each Bt subspecies and isolates (Frankenhuyzen, 2009).

Cry proteins are produced as protoxins that are proteolytically converted into a combination of up to four smaller toxins upon ingestion. These proteins bind to specific receptors in the larval midgut epithelium causing the formation of large cation-selective pores that increase the water permeability of the cell membrane. A large uptake of water then causes cell swelling and rupture of the midgut. Poisoned insects can die quickly from the toxin activity or may die within 2-3 days from septicemia due to the entering of gut contents into the bloodstream. Bt strains containing mixtures of up to 6-8 Cry proteins have been used as microbial pesticides since Bt var. kurstaki have been commercially available since 1961 (Montesinos, 2003). Formulations are active against insect order Lepidoptera (moths and butterflies); Diptera (flies and mosquitoes); Coleoptera (beetles and weevils) and Hymenoptera (bee and wasps) larvae (Frankenhuyzen, 2009). Of the recognized subspecies of Bt, var. kurstaki is toxic to gypsymoth, cabbage looper, and caterpillars (order Lepidoptera), var. israelensis is toxic to fungus gnat larvae, mosquitoes (species of *Aedes* and *Psorophora*), coffee berry borer (Mendez-Lopez et al., 2003), back fly, and some midges (order Diptera), var. san diego is effective against potato beetle, elm leaf beetle, and boll weevils (Whalon & McGaughey, 1998), var. aizawai is effective against wax moth larvae and diamondback moth caterpillar and var. morrisoni is toxic against moth and butterfly caterpillars (order Lepidoptera) (Chattopadhyay et al., 2004). A number of Bt-derived products were used in Europe to control Lepidoptera pests in vegetables, tomatoes, top fruit, vines, olives and forestry (Butt et al., 1999).

Besides Cry proteins (crystal delta endotoxins), Cyt proteins (cytolysins) have been described as another class of insecticidal protein produced by Bt (Yokoyama et al., 1998). Cytolysins interact with phospholipid receptors on the cell membrane in a detergent-like manner (Gill et al., 1987). The hydrophobic portion of the cytolysins bind the amphipathic phospholipids; transmembrane pores are formed and cells are lysed by osmotic lysis (Knowles & Ellar, 1987). The spore inclusions contain many proteins, which sometimes possess distinct activities and may act in a synergistic manner (Yokoyama et al., 1998).

With regard to toxicity, Cry proteins are non toxic to vertebrate species even at doses higher than 1×10^6 µg / kg body weight, while dosages acutely toxic to susceptible insects are about µg/kg body weight (Rosell et al., 2008), however Bt formulations can cause skin and eye irritation (Siegel & Shadduck, 1990). The acidic environment of the mammalian stomach does not favor solubilization and activation of the Cry proteins. These proteins are degraded very fast (often in some seconds), from 60–130 kDa to polypeptides less than 2 kDa that corresponds to peptides with 10 amino acids in length. The rapid degradation of these proteins by proteases in the mammalian gastrointestinal tract precludes their toxicity in mammals. Several studies in vertebrates have failed to find high affinity Cry protein binding sites on gut epithelial cell membranes (Rosell et al., 2008). Bt has thus become a bioinsecticide of great agronomical importance and is classified as toxicity class III pesticide (slightly toxic).

The commercial Bt products are powders comprised of a mixture of dried spores and toxin crystal proteins and these are applied to areas like leaves and roots where insects feed. The commercial Bt product contains about 2.5×10^{11} viable spores per gram. Bt products are marketed worldwide and they account for about 1% of the total agrochemical market. Bt products are known to lose their effectiveness to some extent when stored for longer than six months (Joshi, 2006).

Other species of *Bacillus*, including *B. sphaericus*, *B. popilliae*, *B. subtilis*, *B. lentimorbus*, *B. pumilus* and *B. firmus* have been applied as biopesticides (Schisler et al., 2004).

Bacteria belonging to other genera such as *Pseudomonas fluorescens*, *P. syringae*, *P. putida*, *P. chlororaphis*, *Burkholderia cepacia* and *Streptomyces griseoviridis* have also been used as biopesticides (Montesinos, 2003). However, these bacteria generally lose viability when stored for several weeks, a disadvantage when compared with spore-forming *Bacillus* that demonstrates better shelf-life and facilitates the development of commercial products.

Insect resistance to Bt toxins has led to pursue suitable alternatives. Two more bacteria that are also known to produce insecticidal toxins are *Xenorhabdus* and *Photorhabdus* (both of these belong to the family *Enterobacteriaceae*). Both bacteria are entomophatogens, *Xenorhabdus luminescens* is found to occur in a specialized intestinal vesicle of the nematode *Steinernema carpocapsae* (Akhurst & Dunphy, 1993) with which it maintains a symbiotic relationship. *Photorhabdus luminescens* maintains a symbiotic relationship with nematodes of the family Heterorhabditidae (Poinar, 1990) and is present throughout the intestinal tract of these nematodes. In both mutualistic associations, the nematodes and the bacteria complement each other: the nematode acts as a vector and transports the bacteria into the target insect larva where it bores holes in the intestinal walls of the insect and releases the bacteria in the hemolymph. In the absence of the nematode, the bacteria cannot penetrate into the hemocoel (Tanada & Kaya, 1993). Both the nematode and the bacteria release insecticidal toxins, which eventually kill the insect (Poinar et al., 1977). The bacteria causes septicemia in the insect, the insect is killed and its tissues are used as nutrients (Kaya & Gaugler, 1993). Moreover, bacteria are required by the nematodes for their development into the infective juvenile stage and thus are required for efficient completion of the nematode life cycle. In the absence of the bacterium the nematode cannot reproduce (Tanada & Kaya,1993). With emerging resistance to Bt among insects, *Xenorhabdus* and *Photorhabdus* are considered the next generation of microbial insecticides.

2.4.1.2 Fungi

Fungi often act as important natural control agents against insects, pathogenic fungi, nematodes and as herbicide. Many fungi utilized as biopesticides are pathogenic to insect hosts, therefore they are referred as entomopathogenic fungi; among them, members of Entomophtorales (Zygomycota) and Hyphomycetes are currently under research (Srivastava et al., 2009). Fungal strains are considered suitable for biopesticide development because, unlike other microorganisms, the infectious propagules (conidia) do not need to be ingested and contact with cuticle permits the fungi to penetrate the insect body (Thomas & Read, 2007).

Fungi can act as insecticide by two ways:

a. Infection: most of the fungi species cause death to the insect through asexual spores called conidia. The conidium is the infective unit of entomopathogenic fungi and binds to the host cuticle by nonspecific interaction mediated by cuticle degrading enzymes present on the conidia or by fungal lectins. These conidia enter through the body wall of the host pest by dissolving the body wall by the combined action of enzymes, i.e., chitinase and protease, secreted by the fungi. Fungal penetration is further enhanced by mechanical force. The site of invasion is often between the mouth parts, at intersegmental folds or through spiracles, where locally high humidity promotes germination and the cuticle is nonsclerotized and more easily penetrated. Under favorable environmental conditions (>95% humidity) the fungus will break out through the cuticle and sporulate; it may grow profusely in the blood and give the carcass a characteristic mummified appearance. Therefore, insect death is probably the result of obstruction of blood circulation, starvation or physiological/biochemical disruption brought about by the fungus. The whole procedure takes 3–14 days for insect death (Roy et al., 2006).

b. Mycotoxins: another fungi mode can cause death of the host by the production of mycotoxins, which can interfere in the nervous system of insects. Mycotoxins such as aflatoxin B, trichothecenes, patuline and ochratoxin are reported to be toxic to insects, (Figure 16), (Srivastava et al., 2009).

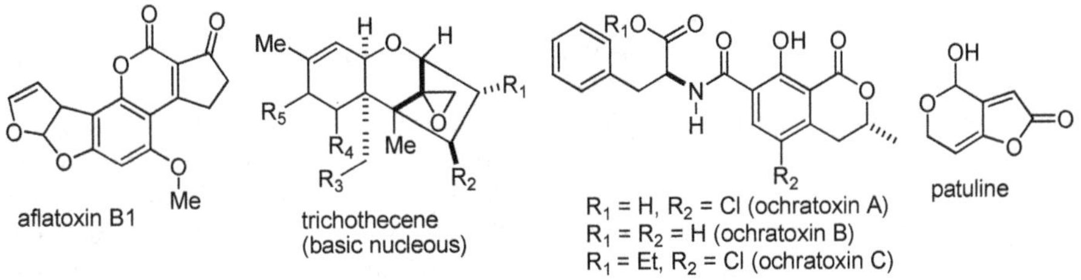

aflatoxin B1   trichothecene R_1 = H, R_2 = Cl (ochratoxin A)   patuline
                           (basic nucleous) R_1 = R_2 = H (ochratoxin B)
                                                     R_1 = Et, R_2 = Cl (ochratoxin C)

Fig. 16. Examples of mycotoxins produced by fungi

Fungi are known to infect a broader range of insects belonging to orders Lepidoptera, Homoptera, Hymenoptera, Coleoptera and Diptera. *Beauveria bassiana, Beauveria brongniari, Metarhizium anisopliae, Metarhizium flavoviride* and *Lagenidium giganteum* are examples of commercially available mycoinsecticides (Rosell et al., 2008).

Trichoderma harzianum, T. viride, Talaromyces flavus, Gliocladium virens , Phytium oligandrum shows fungicide activity against soil-borne pathogenic fungi (Montesinos, 2003). The application of a biopesticide containing the fungus *Verticillium lecanii* was reported to suppress the growth of plant pathogens as well as insect pests (Koike et al., 2005).

A formulation containing the unicellular fungi *Candida oleophila* O is used as a post-harvest biofungicide to control the pathogens *Botrytis cinerea* (gray mold) and *Penicillium expansum* (blue mold) which cause deterioration of apples and pears (Environmental Protection Agency-EPA, 2009).

The main difficulties to be overcome for applying entomophagous fungi in pest control are: scant production of mycotoxins; (ii) carcinogenic mycotoxicosis in non-target organisms; and (iii) slow effectiveness of entomophagous conidia. The combination of fungi formulations with plant extracts exploring their synergistic action is an alternative strategy to overcome these problems (Srivastava et al., 2009).

2.4.1.3 Viruses

Virus-based biopesticides have been used as insect control agents. The larvae of many insect species are vulnerable to viral diseases. Baculoviruses are a large virus group belonging to the family *Baculoviridae* and can infect different insect orders, particularly Lepidoptera and Diptera (Theilmann et al., 2005; Moscardi, 1999). Theilmann et al., 2005; Moscardi, 1999). Baculoviruses are classified into two genera: nuclear polyhedrovirus (NPV) and granulovirus (GV), (Cory & Hails, 1997; McCutchen & Flexner, 1999). Two morphologically distinct forms of infectious particles are generated in the baculovirus cycle, the occlusion derived virus (ODVs), comprising enveloped virions embedded within a crystalline matrix of protein (polyhedrin for NPVs and granulin for GVs), and budded virus (BVs), consisting of a single virion enveloped by a plasma membrane. Due to their specificity and high virulence to a number of insect pest species, they have been used worldwide to control lepidopteran pests in many crops (Moscardi, 1999). BVs are responsible for the systemic or cell-to-cell spread of the virus within an infected insect. OVs, in turn, are responsible for the larva-to-larva transmission of the virus (Inceoglu et al., 2006).

Viruses, like bacteria, must be ingested to infect the insect hosts. During infection the host larvae is debilitated, resulting in reduced movement and increased exposure to predators. Post larval effects include the reduction in reproductive capacity and longevity. Disease and death insects serve as inoculums for virus transmission which may occur by rain and movement of insects on plants (Rosell et al., 2008). The commercial formulations that have been used include Granulosis virus to control *Byctiscus betulae,* Pine sawfly NPV to control *Diprion similis, Heliothis* NPV to control *Helicoverpa zea,* Gypsy moth NPV to control *Lymantria dispar, Mamestria brassicae* NPV to control *Heliothis* (Montesinos, 2003).

Insect viruses are safe to vertebrates, plants and non-target organisms. Limitations on the use of virus formulations include narrow spectrum of biological activity, slow mode of action (5–7 days after ingestion of NPVs and 7–14 days in the case of GV infections), and photolability (solar radiation), (Rosell et al., 2008). The major success of microbial control with viruses takes place in forestry. Forest pests are good targets for viral pesticides because the permanence in forest environment contributes to the pathogen cycle and the forest canopy also helps to protect viral particles from radiation. There have been different approaches directed to enhance the role of baculovirus as effective biopesticides. For instance, the effect of baculovirus may be enhanced by the synergistic action of specific chemical insecticides, such as the pyrethroids deltamethrin and permethrin (McCutchen & Flexner, 1999). To improve the potency and rapid action, recombinant baculovirus have been developed (Bonning & Hammock, 1996).

2.4.1.4 Protozoa

Some protozoan pathogens can kill insect hosts; however, many of them cause chronic infections with debilitating effects (Lacey & Goettel, 1995). One important consequence of protozoan infection is the reduction in the number of offsprings by the infected insects. Species of the genera *Nosema sp.* and *Vairimorpha necatrix* offer the greatest biopesticide potential. *Nosema locustae* is a specie of Microsporidium commercially available to control grasshoppers and crickets. It is most effective when ingested by immature grasshoppers (early nymphal stages). The spore formed by the protozoan is the infection stage in susceptible insects; it germinates in the midgut and causes a slow progress infection where the pathogen causes death three to six weeks after initial infection (Rosell et al., 2008). *Ostrinia nubilalis* that causes important damages to corn was controlled by *Nosema pyrausta* infection, which reduced the egg production per female 53 and 11% at the 16 and 27°C temperature, respectively (Bruck et al., 2001). *Nosema locustae* has been used to reduce grasshopper population in rangeland areas; although not all insects are killed, the infected grasshoppers consume less forage and the females produce fewer eggs. However, the utility of *N. locustae* as biopesticide remains questionable because of the difficulty to determine the treatment efficacy in this highly mobile insect.

2.4.2 Challenges of microbial pesticides

The main problems that should be solved regarding the widespread use of microbial pesticides include their specificity, once they are not effective against a wide range of pests. Although specificity is considered an advantage, it also limits the potential market and increases costs when compared to synthetics. Another important aspect is that biopesticide preparations are sensitive to heat, desiccation and ultraviolet radiation, reducing their effectiveness. Special formulations and storage conditions are necessary; this in turn can complicate the distribution and application of products. Molecular genetics of microorganisms and genetic engineering technology will help in the development of new strategies for biopesticide improvement and its use. More work should be done to enhance shelf-life, to increase the speed of kill, the biological spectrum and the field efficacy of biopesticides.

3. Conclusion

The pollution of the environment by pesticides is a consequence of the continuous agricultural expansion, combined with the population increase. Pesticides are used in sizeable areas and applied to soil surfaces and accumulate beneath the ground surface, reaching rivers and seas. The natural microbiota is continuously exposed to pesticides therefore, it is no surprise that these microorganisms, that inhabit in polluted environments, are armed with resistance by catabolic processes to remove the toxic compounds. Biological degradation by organisms (fungi, bacteria, viruses, protozoa) can efficiently remove pesticides from the environment, especially organochlorines, organophosphates and carbamates used in agriculture. The enzymatic degradation of synthetic pesticides with microorganisms represents the most important strategy for the pollutant removal, in comparison with non-enzymatic processes. Regarding the use of biopesticides, their main advantage is their environmental–friendly nature when compared to chemicals.

To improve the use of microbe–based processes some questions still have to be answered, such as the long term impact of introducing microorganims into the environment, as well as the narrow range of applications (particularly in the case of biopesticides).

The degradation of persistent chemical substances by microorganisms in the natural environment has revealed a larger number of enzymatic reactions with high biorremediation potential. These biocatalysts can be obtained in quantities by recombinant DNA technology, expression of enzymes, or indigenous organisms, which are employed in the field for removing pesticides from polluted areas.

The microorganisms contribute significantly for the removal of toxic pesticides used in agriculture and in the absence of enzymatic reactions many cultivable areas would be impracticable for agriculture.

4. References

Abramowicz, D.A. (1995). Aerobic and Anaerobic PCB Biodegradation in the Environment. *Proceeding from Conference on Biodegradation: Its Role in Reducing Toxicity and Exposure to Environmental Contaminants*, Triangle Park, North Carolina, June, 1995.

Agarwal, H.C.; Singh, D.K. & Sharma, V.B. (1994). Persistence and Binding of *p,p'*-DDE in Soil, *Journal of Environmental Science and Health*, Vol. 29, pp.87-96, ISSN 1093-4529.

Ahrens, R. & Weber, C. (2009). *DDT und die Stockholmer Konvention – Staten am Rande der Legalität.* Pestizid Aktions-Netzwerk (PAN) e.V., ISBN 978-3-9812334-3-8, Hamburg, Germany. ISBN 12-084442-7, New York, USA.

Aislabie, J. & Lloyd-Jones, G. (1995). A Review of Bacterial Degradation of Pesticides, *Australian Journal of Soil Research*, Vol. 33, No. 6, pp.925-942, ISSN 0004-9573.

Aislabie, J.; Richards, N.K. & Boul, H.L. (1997). Microbial Degradation of DDT and its Residues – A Review, *New Zealand Journal of Agricultural Research*, Vol. 40, (January 1997), pp. 269-282, ISSN 0028-8233.

Akhurst, R.J. & Dunphy, G.B. (1993). Tripartite Interactions Between Symbiotically Associated Entomopathogenic Bacteria, Nematodes and Their Insect Hosts, In: *Parasites and Pathogens of Insects*, N. Beckage, S. Thompson & B. Federici, (Eds.) ,Vol. 2, pp. 1-23, Academic Press, ISBN 0-Atlas, R. M. (2nd. Ed). (1990). *Microbiology: Fundamentals and Applications*, MacMillan, ISBN 978-002-3045-50-9, New York, USA.

Arisoy, M. & Kolankaya, N. (1998). Biodegradation of Heptachlor by *Phanerochaete chrysosporium* ME 446: The Toxic Effects of Heptachlor and its Metabolites on Mice, *Turkish Journal of Biology*, Vol. 22, pp.427-434, (May 1997), ISSN 1303-6092.

Barriuso, E.; Benoit, P. & Dubus, I.G. (2008). Formation of Pesticide Nonextractable (Bound) Residues in Soil: Magnitude, Controlling Factors and Reversibility, *Environmental Science & Technology*, Vol.42, No.6, (February 2008), pp. 1845-1854, ISSN 0013-936X.

Benimeli, C.S.; Fuentes, M.S.; Abate, C.M. & Amoroso, M.J. (2008). Bioremediation of Lindane-Contaminated Soil by *Streptomyces* sp. M7 and its Effects on *Zea mays* Growth, *International Biodeterioration & Biodegradation*, Vol. 61, (September 2007), pp.233-239, ISSN 0964-8305.

Benn, F.R. & McAuliffe, C.A. (1975). *Chemistry and Pollution*, Macmillan Press, ISBN 333138880, London, England.

Bollag, J.M. & Liu, S.Y. (1990). Biological Transformation Processes of Pesticides, In: *Pesticides in the Soil Environment Processes, Impacts and Modeling*, CHENG, H.H. (Ed.), pp.169-211, Soil Science Society of America, ISBN 0-89118-791-X, Madison, Wisconsin, USA.

Bondarenko, S.; Gan, J.; Haver, D. L. & Kabashima, J.N. (2004). Persistence of Selected Organophosphate and Carbamate Insecticides in Waters from a Coastal Watershed, *Environmental Toxicology and Chemistry*, Vol.23, No.11, (November 2004), pp. 2649-2654, ISSN 0730-7268.

Bonning, B.C. & Hammock, B. D. (1996). Development of Recombinant Baculoviruses for Insect Control, *Annual Review of Entomology*, Vol. 41, pp.191–210, ISSN 1545-4487.

Boul, K.H.; Garnham, M.L.; Hucker, D.; Baird, D. & Aislabie, J. (1994). The Influences of Agricultural Practices on the Levels of DDT and its Residues in Soil, *Environmental Science and Technology*, Vol. 28, pp.1397-1402, ISSN 1520-5851.

Bruck, D.J.; Lewis, L.C. & Gunnarson, R.D. (2001). Interaction of *Nosema pyrausta* and Temperature on *Ostrinia nubilalis* Egg Production and Hatch, *Journal of Invertebrate Pathology*, Vol.78, No.4, (November 2001), pp. 210–214, ISSN 0022-2011.

Bumpus, J.A.; Powers, R.H. & Sun, T. (1993). Biodegradation of DDE (1,1-Dichloro-2,2-bis(4-hlorophenyl)ethane) by *Phanerochaete chrysosporium*, *Mycological Research*, Vol. 97, pp.95-98, ISSN 0953-7562.

Butt, T.M.; Harris, J.G. & Powell, K.A. (1999). Microbial Biopesticides: the European Scene, In: *Biopesticides Use and Delivery*, D. R. Hall & J.J. Menn, (Eds.), pp. 23–44, Humana Press, ISBN 0-89603-515-8, New Jersey, USA.

Chanika, E.; Georgiadou, D.; Soueref, E.; Karas, P.; Karanasios, E.; Nikolaos, G.T.; Tzortzakakis, E.A. & Karpouzas, D.G. (2011). Isolation of Soil Bacteria Able to Hydrolyze Both Organophosphate and Carbamate Pesticides, *Bioresource Technology*, Vol.102, (February 2011), pp. 3184-3192, ISSN 09608524.

Chattopadhyay, A.; Bhatnagar, N.B. & Bhatnagar, R. (2004). Bacterial Insecticidal Toxins, *Critical Reviews in Microbiology*, Vol.30, No.1, (March 2002), pp.33–54, ISSN 1040-8371.

Chaudhry, G.R. & Chapalamadugu, S. (1991). Biodegradation of Halogenated Organic Compounds Microbiological Reviews, *Microbiology and Molecular Biology Reviews*, Vol.55, No.1, (March 1991), pp. 59-79, ISSN 1092-2172.

Colosio, C.; Tiramani, M.; Brambilla, G.; Colombi, A. & Moretto, A. (2009). Neurobehavioural Effects of Pesticides with Special Focus on Organophosphorus Compounds: Which is the Real Size of the Problem?, *Neurotoxicology*, Vol.30, No.6, (November 2009), pp. 1155-1161, ISSN 0161-813X.

Cory; J.S. & Hails, R.S. (1997). The Ecology and Biosafety of Baculoviruses, *Current Opinion in Biotechnology*, Vol.8, No.3, (June 1997), pp.323–327, ISSN 0958-1669.

Crawford, R.L. & Mohn, W.W. (1985). Microbiological Removal of Pentachlorophenol from Soil Using a *Flavobacterium*, *Enzyme and Microbial Technology*, Vol. 7, No. 12, (December 1985), pp.617-620, ISSN 0141-0229.

D'Amato, C., Torres, J. P. M. & Malm, O. (2002). DDT (Dicloro Difenil Tricloroetano): Toxicidade e Contaminação Ambiental – Uma Revisão, *Química Nova*, Vol. 25, No. 6, pp.995-1002, ISSN 0100-4042.

Debarati, P.; Gunjan, P.; Janmejay, P.; Rakesh,V.J.K. (2005). Accessing Microbial Diversity for Bioremediation and Environmental Restoration, *Trends in Biotechnology*, V.23, No.3, (March 2005), pp.135-142, ISSN 0167-9430.

De Bleecker, J.L. (2008). Organophosphate and Carbamate Poisoning, *Handbook of Clinical Neurology*,Vol.91, pp.401-432, ISSN0072-9752.

De Schrijver, A. & De Mot, R. (1999). Degradation of Pesticides by Actinomycetes, *Critical Review in Microbioloty*, Vol. 25, No. 2, pp. 85-119, ISSN 1040-841X.

Diaz, E. (2004). Bacterial Degradation of Aromatic Pollutants: A Paradigm of Metabolic Versatility, *International Microbiology*, Vol. 7, No. 3, (September 2004), pp. 173-180, ISSN 1139-6709.

Dua, M.; Singh ,A.; Sethunathan, N. & Johri, A.K. (2002). Biotechnology and Bioremediation: Successes and Limitations, *Applied Microbiology and Biotechnology*, Vol.59, No.2-3, (February 2002), pp. 143-152, ISSN 0175-7598.

EPA New Biopesticides Active Ingredients (2009), 07.02.2011, Available from http://www.epa.gov/oppbppd1/biopesticides/product_lists/new_ai_2009.html.

Fang, H.; Dong, B.; Yan, H.; Tang, F. & Yunlong, Y. (2010). Characterization of a Bacterial Strain Capable of Degrading DDT Congeners and its Use in Bioremediation of Contaminated Soil, *Journal of Hazardous Material*, Vol. 184, Nos.1-3, (August 2010), pp.281-289, ISSN 0304-3894.

Finley, S.D.; Broadbelt, L.J. & Hatzimanikatis, V. (2010). In Silico Feasibility of Novel Biodegradation Pathways for 1,2,4-Trichlorobenzene, *BMC Systems Biology*, Vol.4, No.7, (February 2010), pp.4-14, ISSN 1752-0509.

Frankenhuyzen, K.V. (2009). Insecticidal Activity of *Bacillus thuringiensis* Crystal Proteins, *Journal of Invertebrate Pathology*, Vol.101, No.1, (April 2009), pp. 1–16, ISSN 0022-2011.

Franzmann, P.D.; Zappia, L.R.; Tilbury, A.L.; Patterson, B.M.; Davis, G.B.; Mandelbaum, R.T. (2000). Bioaugmentation of Atrazine and Fenamiphos Impacted Groundwater:

Laboratory Evaluation, *Bioremediation Journal,* Vol.24, No.3, pp.48-68, ISSN 10889868.

Fravel, D.R. (2005). Commercialization and Implementation of Biocontrol, Annual Review of Phytopathology, Vol.43, (July 2005), pp. 337-359, ISSN 0066-4286.

Fries, G.R.; Marrow, G.S. & Gordon, C.H. (1969). Metabolism of *o,p'*-DDT by Rumen Microorganisms, *Journal of Agricultural and Food Chemistry,* Vol. 17, No. 4, pp.860-862, ISSN 1520-5118

Gavrilescu, M. (2005). Fate of Pesticides in the Environment and its Bioremediation, *Engineer in Life Science,* Vol.5, No. 6, (December 2005), pp. 497-526, ISSN 1618-2863.

Gerhardson, B. (2002). Biological Substitutes for Pesticides, *Trends in Biotechnology,* Vol. 20, No. 8, (August 2002), pp. 338 343, ISSN 0167-9430.

Gill, S.S.; Singh, G.J.P. & Hornung, J.M. (1987). Cell Membrane Interaction of *Bacillus thuringiensis* Subsp. Israelensis Cytolytic Toxins, *Infection and Immunity,* Vol.55, No.5 (May 1987), pp.1300–1308, ISSN 0019-9567.

Ghosh, P.G.; Sawant, N.A.; Patil S.N.; Aglave, B.A. (2010). Microbial Biodegradation of Organophosphate Pesticides, *International Journal of Biotechnology and Biochemistry,* Vol.6, No.6, pp.871-876, ISSN 0973-2691.

Gonzales, J.M. & Carlton, B.C. (1980). Patterns of Plasmid DNA in Crystalliferous and Acrystalliferous Strains of BT, *Plasmid,* Vol.3, No.1, pp.92-98, ISSN 0147-619X.

Griza, F.T.; Ortiz, K.S.; Geremias, G; Thiesen, F.V. (2008). Avaliação da Contaminação por Organofosforados em Águas Superficiais no Município de Rondinha- Rio Grande do Sul, *Quimica Nova,* Vol.31, No.7, pp.1631-1635, ISSN 0100-4042.

Guenzi, W.D. & Beard, W.E. (1967). Anaerobic Biodegration of DDT to DDD in Soil, *Science.* Vol. 156, No.3778, (May 1967), pp.1116-1117, ISBN 0036-8075.

Gupta, R. C. (2006). Toxicology of Organophosphate & Carbamates Compounds, Academic Press, ISBN 10: 0-12-088523-9.

Hemmert, A.C. & Redinbo, M.R. (2010). A Structural Examination of Agrochemical Processing by Human Carboxylesterase 1, *Journal of Pesticides Science,* Vol.35, No.3, (June 2010), pp.250–256, ISSN 1348-589X.

Hong, L., Zhang, J.J., Wang, S.J., Zhang, X.E., Zhou, N.Y. (2005). Plasmid-Borne Catabolism of Methyl Parathion and *p*-Nitrophenol in *Pseudomonas* sp. Strain WBC-3, *Biochemistry and Biophysic Research Communications,* Vol.334, No. 4, (September 2005), pp.1107-1114, ISSN 0006-291X.

Horne, I.; Sutherland, T.D.; Harcourt, R.L.; Russell, R.J.; Oakeshott, J.G. (2002). Identification of an *opd* (Organophosphate Degradation) Gene in an *Agrobacterium* Isolate, *Applied and Environmental and Microbiology,* Vol.68, No.7, (July 2002), pp. 3371–3376, ISSN 0099-2240.

Hussain, S.; Siddique, T.; Arshad, M. & Saleem, M. (2009). Bioremediation and Phytoremediation of Pesticides: Recent Advances, *Critical Review in Environmental Science and Technology,* Vol. 39, No. 10, pp. 843-907, ISSN 1064-3389.

Inceoglu, A.B.; Kamita, S.G. & Hammock, B.D. (2006). Genetically Modified Baculoviruses: A Historical Overview and Future Outlook, *Advances in Virus Research,* Vol. 68, pp.323–360, ISBN 978-0-12-039868-3.

Jacobsen, O.S. & Bossi, R. (1997). Degradation of Ethylenethiourea (ETU) in Oxic and Anoxic Sandy Aquifers, *FEMS Microbiology Review,* Vol. 20, No. 3-4, (July 1997), pp. 539-544, ISSN 0168-6445.

Jain, R.K.; Kapur, M.; Labana, S.; Lal, B.; Sarma, P.M.; Bhattacharya, D.; Thakur, I.S. (2005). Microbial Diversity: Application of Microorganisms for the Biodegradation of Xenobiotics, *Current Science,* Vol.89, No.1, (July 2005), pp.101-112, ISSN 0011-3891.

Johnsen, R.E. (1976). DDT Metabolism in Microbial Systems, *Pesticide Reviews,* Vol. 61, pp.1-28.

Jokanovic, M. & Prostran M. (2009). Pyridinium Oximes as Cholinesterase Reactivators. Structure-Activity Relationship and Efficacy in the Treatment of Poisoning with Organophosphorus Compounds, *Current Medicinal Chemistry,* Vol.16, No.17, pp. 2177-2188, ISSN 0929-8673.

Joshi, S.R. (2006). *Biopesticides: A Biotechnological Approach,* New Age International Publishers, ISBN 81-224-1781-7, New Delhi, India.

Kamanavalli, C.M. & Ninnekar, H.Z. (2005). Biodegradation of DDT by a *Pseudomonas* Species, *Current Microbiology,* Vol. 48, No. 1, (March 2005), pp.10-13, ISSN 0343-8651.

Kannan, K.; Tanabe, S.; Willians, R.J. & Tatsukawa, R. (1994). Persistent Organochlorine Residues in Foodstuffs from Australia, Papua New Guinea and the Solomon Islands: Contamination Levels and Dietary Exposure, *Science of the Total Environment,* Vol. 153, pp.29-49, ISSN 0048-9697.

Karpouzas, D.G. & Singh, B.K. (2006). Microbial Degradation of Organophosphorus Xenobiotics: Metabolic Pathways and Molecular Basis, *Advances in Microbial and Physiology,* Vol.15, No.51, pp.119-225, ISSN 0065-2911.

Kaufman, D.D. & Blake, J. (1973). Microbial Degradation of Several Acetamide,Acylanilide, Carbamate, Toluidine and Urea Pesticides. *Soil Biology & Biochemistry,* Vol.5, No. 3, (November 1972), pp. 297-308, ISSN 0038-0717.

Kaya, H.K. & Gaugler, R. (1993). Entomopathogenic Nematodes, *Annual Review of Entomology,* Vol.38, pp. 181-206, ISSN 1545-4487.

Kim, I.S.; Ryu, J.Y.; Hur, H.G.; Gu, M.B.; Kim, S.D. & Shim, J.H. (2004). *Sphingomonas* sp. Strain SB5 Degrades Carbofuran to a New Metabolite by Hydrolysis at the Furanyl Ring, *Journal of Agricultural and Food Chemistry,* Vol. 52, No. 8, (April 2004), 2309-2314, ISSN 0021-8561.

Knowles, B.H. & Ellar, D. (1987). Colloid-osmotic Lysis is a General Feature of the Mechanism of Action of *Bacillus thuringiensis* δ-Endotoxins with Different Insect Specificity, *Biochimica et Biophysica Acta,* Vol.924, No.3, (June 1987), pp. 509–518, ISSN 0304-4165.

Koike, M.H.; Yoshida, S.I.; Abe, N. & Asano, K.M. (2005). *Microbial Pesticide Inhibiting the Outbreak of Plant Disease Damage.* Patent application, WO 2005/104853. Date of filing: 27.04.2005, Date of publication: 18.04.2007.

Konradsen, F.; Van der Hoek, W.; Amerasinghe, F. P.; Mutero, C. & Boelee, E. (2004). Engineering and Malaria Control: Learning from the Past 100 Years, *Acta Tropica,* Vol. 89, No.2, (January 2004), pp.99-108, ISSN 0001-706X.

Lacayo, R.M., Terrazas, E., van Bavel, B. & Mattiasson, B. (2006). Degradation of Toxaphene by *Bjerkandera* sp. Strain BOL13 Using Waste Biomass as a Co-substrate, *Applied*

Microbiology and Biotechnology, Vol. 71, No.4, (May 2005), pp.549-554, ISSN 0175-7598.

Lacey, L.A. & Goettel, M.S. (1995). Current Developments in Microbial Control of Insect Pests and Prospects for the Early 21st Century, *Entomophaga* ,Vol.40, pp.3–27, ISSN 0013-8959.

Lal, R. & Saxena, D.M. (1982). Accumulation, Metabolism, and Effects of Organochlorine Insecticides on Microorganisms, *Microbiological reviews*, Vol. 46, No. 1, (March 1982), pp.95-127, ISSN 0146-0749.

Lan, W.S.; Gu, J.D.; Zhang, J.L.; Shen, B.C.; Jiang, H.; Mulchandani, A.; Chen, W. & Qiao C.L. (2006). Coexpression of Two Detoxifying Pesticide-degrading Enzymes in a Genetically Engineered Bacterium, *International Biodeterioration and Biodegradation*, Vol. 58, No.2, (September 2006), pp. 70-76, ISSN 0964-8305.

Langlois, B.E.; Collins, J.A. & Sides, K.G. (1970). Some Factors Affecting Degradation of Organochlorine Pesticide by Bacteria, *Journal of Dairy Science*, Vol.53, No. 12, (December 1970), pp. 1671-1675, ISSN 0022-0302.

Liu, Z.; Hong, Q.; Xu, J.H.; Wu, J.; Zhang, X.Z.; Zhang, X.H.; Ma, A.Z.; Zhu, J.; Li, S.P. (2003). Cloning, Analysis and Fusion Expression of Methyl Parathion Hydrolase, *Acta Genetica Sinica*, Vol.30, No.11, pp.1020-1026, ISSN 1671-4083.

Mariconi, F.A.M. (7th. edition). (1985). *Inseticidas e Seu Emprego no Combate às Pragas*, Editora São Paulo, São Paulo, Brazil.

Masse, R.; Lalanne, D.; Meisser, F. & Sylvester, M. (1989). Characterization of New Bacterial Transformation Products of 1,1,1-Trichloro-2,2-bis-(4-chlorophenyl)ethane (DDT) by Gas Chromatography/Mass Spectrometry, *Biomedical and Environmental Mass Spectrometry*, Vol. 18, No. 9, pp.741-752, ISSN 0887-6134.

Matolcsy, G.; Nádasy, M. & Andriska, V. (1988). *Pesticide Chemistry*, Elsevier Science Publishing Co., ISBN 978-044-4989-03-1, New York, USA.

Matsumura, F.; Boush, G. M. & Tai, A. (1968). Breakdown of Dieldrin in the Soil by a Microorganism, *Nature*, Vol. 219, No. 5157, (August 1968), pp.965-967, ISSN 0028-0836.

McCutchen, W. F. & Flexner, L. (1999). Join Action of Baculovirus and Other Control Agents, In: *Biopesticides Use and Delivery*, D. R. Hall & J.J. Menn, (Eds.), pp. 341–355, Humana Press, ISBN 0-89603-515-8, New Jersey, USA.

Méndez-López, I.; Basurto-Rios, R. & Ibarra, J.E. (2003). *Bacillus thuringiensis* Serovar Israelensis is Highly Toxic to the Coffee Berry Borer, *Hypothenemus hampei* Ferr (Coleoptera: Scolytidae). *FEMS Microbiology Letters*, Vol. 226, No.1, (September 2003), pp. 73–77, ISSN 1574-6968.

Menone, M.L.; Bortolus, A.; Aizpún de Moreno, J.E.; Moreno, V.J.; Lanfranchi, A. L.; Metcalfe, T.L. & Metcalfe, C.D. (2001). Organochlorine Pesticides and PCBs in a Southern Atlantic Coastal Lagoon Watershed, Argentina, *Archives of Environmental Contamination and Toxicology*, Vol. 40, No.3 (April 2001), pp.355-362, ISSN 0090-4341.

Mishra, V.; Lal, R.; Srinivasan, S. (2001). Enzymes and Operons Mediating Xenobiotic Degradation in Bacteria, *Critical Reviews in Microbiology*,Vol.27,No.2, pp.133-166, ISSN 1040-841x.

Mitra, J. & Raghu, K. (1988). Influence of Green Manuring on the Persistence of DDT in Soil, *Environmental Technology Letters*, Vol. 9, No. 8, pp.847-852, ISSN 0143-2060.

Montesinos, E. (2003). Development, Registration and Commercialization of Microbial Pesticides for Plant Protection, *International Microbiology*, Vol. 6, No. 4, (July 2003), pp. 245-252, ISSN 1139-6709.

Moreby, S.J.; Southway, S.; Barker, A.; Holland, J.M. (2001). A Comparison of the Effect of New and Established Insecticides on Nontarget Invertebrates of Winter Wheat Fields, *Environmental Toxicology and Chemistry*, Vol.20, No.10, (October 2001), pp.2243-2254, ISSN 0730-7268.

Moscardi, F. (1999). Assessment of the Application of Baculoviruses for Control of Lepidoptera, *Annual Review of Entomology*, Vol.44, pp. 257–289, ISSN 1545-4487. New York, USA.

Mulbry, W.W.& Karns, J.S. (1989). Parathion Hydrolase Specified by the *Flavobacterium opd* Gene: Relationship Between the Gene and Protein, *Journal of Bacteriology*, Vol.171, No.12, (December 1989), pp.6740-6746, ISSN 0021-9193.

Novák, J.; Vlasáková, V.; Tykva, R. & Ruml, T. (2003). Degradation of Juvenile Hormone Analog by Soil Microbial Isolates, *Chemosphere* Vol.52, No.1, (July 2003), pp.151-159, ISSN 0045-6535.

Oga, S. *Fundamentos de toxicologia*, 2a ed., Atheneu Editora, São Paulo, 2003.

Oliveira-Silva, J.J.; Alves, S.R.; Meyer, A.; Perez, F.; Sarcinelli, P.N.; Mattos, R.C.O.C. & Moreira, J,C. (2001). Influence of Social-economic Factors on the Pesticine Poisoning, Brazil, *Revista de Saúde Pública*, Vol.35, No.2, (February 2001), pp.130-135, ISSN 0034-8910.

Ortega, N.O., Nitschke, M., Mouad, A.M., Landgraf, M.D., Rezende, M.O.O., Seleghim, M.H.R., Sette, L. D., Porto, A.L.M. (2011). Isolation of Brazilian Marine Fungi Capable of Growing on DDD Pesticide, *Biodegradation*, Vol. 22, (June 2010), pp. 43-50, ISSN 0964-8305.

Ottaway, J.H. (1982). *The Biochemistry of Pollution*, Edward Arnold, ISBN 0-7131-2784-8, London, England.

Patil, K.C.; Matsumura, F. & Boush, G.M. (1970). Degradation of Endrin, Aldrin, and DDT by Soil Microorganisms, *Journal of Applied Microbiology*, Vol. 19, No. 5, (May 1970), pp.879-881, ISSN 1365-2672.

Patnaik, P. (2003). *A Comprehensive Guide to the Hazardous Properties of Chemical Substances*, John Wiley & Sons, ISBN 978-0-471-71458-3, New Jersey, USA.

Pattanasupong, A.; Nagase, H.; Sugimoto, E.; Hori, Y.; Hirata, K.; Tani, K.; Nasu, M. & Miyamoto, K. (2004). Degradation of Carbendazim and 2,4-Dichlorophenoxyacetic Acid by Immobilized Consortium on Loofa Sponge, *Journal of Bioscience and Bioengineering*, Vol. 98, No.1, pp. 28-33, ISSN 1389-1723.

Pesce, S.F. & Wunderlin, D.A. (2004). Biodegradation of Lindane by a Native Bacterial Consortium Isolated from Contaminated River Sediment. *International Biodeterioration & Biodegradation*, Vol. 54, No.4, (January 2004), pp. 255-260, ISSN 0964-8305.

Pfaender, F.K. & Alexander, M. (1972). Extensive Microbial Degradation of DDT *in vitro* and DDT Metabolism by Natural Communities, *Journal of Agricultural and Food Chemistry*, Vol. 20, No.4, (July 1972), pp.842-846, ISSN 1520-5118.

Poinar, G.O., Thomas, G.M. & Hess, R. (1977). Characteristics of the Specific Bacterium Associated with Heterorhabditis Bacteriophora (Heterorhabditidae: Rhabditida), *Nematologica*, Vol.23, No.1, pp.97-102, ISSN 0028-2596.

Poinar, G. (1990). Biology and Taxonomy of Steinernematidae and Heterorhabditidae. In: *Entomopathogenic Nematodes in Biological Control*, R.R. Gaugler & R.K. Kaya, (Eds.), pp. 23–62 , CRC Press, ISBN 0849345413, Boca Raton, USA.

Raaijmakers, J.M.; Vlami, M. & De Souza, J.T. (2002). Antibiotic Production by Bacterial Biocontrol Agents, *Antonie van Leeuwenhoek*, Vol. 81, Nos. 1-4, (December 2002), pp.537-547, ISSN 0003-6072.

Ragnarsdottir, K. (2000). Environmental Fate and Toxicology of Organophosphate Pesticides, *Journal of the Geological Society London*, Vol. 157, No.4, (July 2000), pp. 859-876, ISSN 0016-7649.

Reed, W.D. & Fukuto, T.R. (1973). The Reactivation of Carbamate-inhibited Cholinesterase, Kinetic Parameters, *Pesticide Biochemistry and Physiology*, Vol.3, No.2, (January 1973), pp.120-130, ISSN 0048-3575.

Rigas, F.; Dritsa, V.; Marchant, R.; Papadopoulou, K.; Avramides, E.J. & Hatzianestis, I. (2005). Biodegradation of Lindane by *Pelourotus ostreatus* via Central Composite Design, *Environmental International*, Vol. 31, No.2, (December 2004), pp.191-196, ISSN 0160-4120.

Rochkind-Dubinsky, M.L.; Sayler, G.S. & Blackburn, J.W. (1987). Microbiological Decomposition of Chlorinated Aromatic Compounds. pp. 153-162. In: *Microbiology series*, Vol. 18, New York, USA. ISSN 0092-6027.

Rosell,G.; Quero, C.; Coll, J. & Guerrero, A. (2008). Biorational Insecticides in Pest Management, *Journal of Pesticide Science*, Vol. 33, No. 2, pp. 103-121, ISSN1348-589X.

Rosman, Y.; Makarovsky, I.; Bentur, Y.; Shrot, S.; Dushnistky, T. & Krivoy, A. (2009). Carbamate Poisoning: Treatment Recommendations in the Setting of a Mass Casualties Event, *American the Journal Emergency of Medicine*, Vol. 27, No.9, (November 2009), pp.1117-1124, ISSN 0735-6757.

Roy, H.E.; Steinkraus, D.C.; Eilenberg, J.; Hajek, A.E. & Pell, J.K. (2006). Bizarre Interactions and Endgames: Entomopathogenic Fungi and their Arthropod Hosts, *Annual Review of Entomology*, Vol.51, (September 2005), pp. 331–357, ISSN 1545-4487.

Sakai, K. & Matsumura, F. (1971). Degradation of Certain Organophosphate and Carbamate Insecticides by Human Brain Esterases, *Toxicology and Applied Pharmacology*, Vol.19, No.4, (August 1971), pp.660-666, ISSN 0041-008X.

Schisler, D. A.; Slininger, P.J.; Behle, R.W. & Jackson, M.A. (2004). Formulation of *Bacillus* spp. for Biological Control of Plant Diseases, *Phytopathology*, Vol.94, No.11, (June 2004), pp.1267–1271, ISSN 0031-949X.

Schoefs, O.; Perrier, M. & Samson, R. (2004). Estimation of Contaminant Depletion in Unsaturated Soils Using a Reduced-order Biodegradation Model and Carbon Dioxide Measurement, *Applied Microbiology and Biotechnology*, Vol.64, No1, (March 2004), pp. 53-61, ISSN 0175-7598.

Schroll, R.; Brahushi, R.; Dorfler, U.; Kuhn, S.; Fekete, J.; Munch, J.C. (2004). Biomineralisation of 1,2,4-Trichlorobenzene in Soils by an Adapted Microbial Population, *Environmental Pollution*, Vol. 127, No.3, (February 2004), pp.395-401, ISSN 0269-7491.

Scott, H.D. (1st. edition). (2000). *Soil Physics: Agricultural and Environmental Applications*, Iowa State University Press, ISBN 0-8138 — 2087-1, Ames, Iowa, USA.

Serdar, C.M.; Murdock, D.C.; Rohde, M.F. (1989). Parathion Hydrolase Gene from *Pseudomonas diminuta* MG: Subcloning, Complete Nucleotide Sequence, and Expression of the Mature Portion of the Enzyme in *Escherichia coli*, *Nature Biotechnology*, Vol.7, pp.1151-1155, ISSN 1087-0156.

Siegel, J.P. & Shadduck, J.A.(1990). Clearence of *B. sphaericus* and *B. thuringiensis* spp. Israelensis from Mammals, *Jounal of Economic Entomology*, Vol. 83, No.2, pp.347–355, ISSN 0022-0493.

Silva, F.C.; Cardeal, Z.L. & De Carvalho, C.R. (1999). Determination of Organophosphorus Pesticides in Water Using SPME-GC-MS, *Química Nova*, Vol.22, No.2, (September 1998), pp. 197-200, ISSN 0100-4042.

Singh, B.K. & Walker, A. (2006). Microbial Degradation of Organophosphorus Compounds, *FEMS Microbiology Review*, Vol.30, No. 3, (May 2006), pp.428-471, ISSN 0168-6445.

Sogorb, M.A. & Vilanova, E. (2002). Enzymes Involved in the Detoxification of Organophosphorus, Carbamate and Pyrethroid Insecticides Through Hydrolysis, *Toxicology Letters*, Vol. 128, Nos. 1-3, (March 2002), pp. 215-228, ISSN 0378-4274.

Soriano, J.M.; Jiménez, B.; Font, G. & Moltó, J.C. (2001). Analysis of Carbamate Pesticides and Their Metabolites in Water by Solid Phase Extraction and Liquid Chromatography: A Review, Critical Review Analytical Chemistry, Vol.31,No.1, (January 2001), pp.19-52, ISSN 1040-8347.

Spain, J.C. & Nishino, S.F. (1987). Degradation of 1,4-Dichlorobenzene by a *Pseudomonas* sp., *Applied and Environmental Microbiology*, Vol. 53, No. 5, (May 1987), pp.1010-1019, ISBN 0099-2240.

Srivastava, C. N.; Maurya, P.; Sharma, P. & Mohan, L. (2009). Prospective Role of Insecticides of Fungal Origin: Review, *Entomological Research*, Vol.39, No.6, (November 2009), pp. 341–355, ISSN 1748-5967.

Stanlake, G.J. & Finn, R.K. (1982). Isolation and Characterization of a Pentachlorophenol-Degrading Bacterium, *Applied Environmental Microbiology*, Vol. 44, (No. 6, December 1982), pp.1421-1427, ISSN 0099-2240.

Subba-Rao, R.V. & Alexander, M. (1985). Bacterial and Fungal Cometabolism of 1,1,1-Trichloro-2,2-bis(*p*-chlorophenyl)ethane (DDT) and its Breakdown Products, *Applied Environmental Microbiology*, Vol. 49, No.3 (March 1985), pp.509-516, ISSN 1098-5336.

Suzuki, O. & Watanabe, K. (2005). *Drugs and Poisons in Humans - A Handbook of Practical Analysis*. In: Carbamate Pesticides. 559-570, DOI: 10.1007/3-540-27579-7-62. Springer.

Tanada, Y. & Kaya, H.K. (1993). *Insect Pathology*, Academic Press, ISBN 0-12-683255-2.

Theilmann, D.A.; Blissard, G.W.; Bonning, B.; Jehle, J.A.; O'Reilly, D.R.; Rohrmann, G.F.; Thiem, S. &Vlak, J.M. (2005). Baculoviridae, In: *Virus Taxonomy Classification and Nomenclature of Viruses: Eighth Report of the International Committee on the Taxonomy of Viruses*, C.M Fauquet; M.A. Mayo; J. Maniloff; U. Desselberger & L.A. Ball, (Eds.), pp. 177–185, Elsevier, ISBN 0-12-249951-4, New York, USA.

Thomas, M.B. & Read, A.F. (2007). Can Fungal Biopesticides Control Malaria?, *Nature Reviews Microbiology*, Vol.5, No.5, (May 2007), pp. 377–383, ISSN 1740-1526.

Turk, J. (3rd. edition). (1989). *Introduction to Environmental Studies*, Saunders College Publishing, ISBN 9780030114687, New York, USA.

Vaccari, D.A.; Strom, P.F. & Alleman, J,E. (2006). Environmental Biology for Engineers and Scientists. John Wiley&Sons, DOI: 10.1002/0471741795, ISBN-10: 0471722391.

Vallaeys, T.; Fulthorpe, R.R.; Wright, A.M.; Soulas, G. (1996). The Metabolic Pathway of 2,4-Dichlorophenoxycetic Acid Degradation Involves Different Families of tfdA and tfdB Genes According to PCR-RFLP Analysis, *FEMS Microbiology Ecology*, Vol.20, No.3, (July 1996), pp.163-172, ISSN 0168-6496.

Xiao, P., Mori, T., Kamei, I. & Kondo, R. (2011). Metabolism of Organochlorine Pesticide Heptachlor and its Metabolite Heptachlor Epoxide by White-rot Fungi, Belonging to Genus *Phlebia, Microbiology Letters*, Vol. 314, No.2, (January 2011), pp.140-146, ISSN 1574-6968.

Xu, B.; Jianying, G.; Yongxi, Z. & Haibo, L. (1994). Behaviour of DDT in Chinese Tropical Soils, *Journal of Environmental Science and Health Part B*, Vol. B29, No.1, pp.37-46, ISSN 1532-4109.

Wedemeyer, G. (1966). Dechlorination of DDT by *Aerobacter aerogenes, Science*, Vol. 152, No. 3722, (April 1966), pp.647-647, ISSN 0036-8075.

Wedemeyer, G. (1967). Dechlorination of 1,1,1-Trichloro-2,2-bis(p-chlorophenyl)ethane by *Aerobacter aerogene, Journal of Applied Microbiology*, Vol. 15, No. 3, (May 1967), pp.569-574, ISSN 1365-2672.

Whalon, M.E. & McGaughey,W.H. (1998). BT: Use and Resistance Management, In: *Insecticides With Novel Modes of Action, Mechanism and Application*, I. Ishaaya & D. Degheele, (Eds.), pp.106-137, Springer, ISBN 3-5406-3058-9, Berlin, Germany.

WHO (World Health Organization). (2009). *The WHO Recommended Classification of Pesticides by Hazard and Guidelines to Classification: 2009*, Wissenchaftliche Verlagsgesellschaft mbH, ISBN 978-92-4-154796-3, Geneva, Switzerland.

Wolfe, N.L.; Zepp, R.G. & Paris, D.F. (1978). Use of Structure-reactivity Relationships to Estimate Hydrolytic Persistence of Carbamate Pesticides, *Water Research*, Vol.12, No. 8, 561-563, ISSN 0043-1354.

Yair, S.; Ofer, B.; Arik, E.; Shai, S.; Yossi, R.; Tzvika, D. & Amir, K. (2008). Organophosphate Degrading Microorganisms and Enzymes as Biocatalysts in Environmental and Personal Decontamination, *Applied Critical Review in Biotechnology*, Vol.28, pp. 265-275, ISSN 0273-2289.

Yokoyama, Y.; Kohda, K. & Okamoto, M. (1998). CytA Protein, a Dendotoxin of *Bacillus thuringiensis* subsp. Israelensis is Associated with DNA, *Biological & Pharmaceutical Bulletin*, Vol.21, No.12, pp.1263–1266, ISSN 0918-6158.

Yuanfan, H; Jin, Z.; Qing, H.; Qian, W.; Jiandong, J.; Shunpeng, L. (2010). Characterization of a Fenpropathrin-Degrading Strain and Construction of a Genetically Engineered Microorganism for Simultaneous Degradation of Methyl Parathion and Fenpropathrin, *Journal of Environmental Management*, Vol.91, No.11 (November 2010), pp.2295-2300, ISSN 0301-4797.

Zhang, H.; Yang, C.; Li, C.; Li, L.; Zhao, Q.; Qiao, C. (2008). Functional Assembly of a Microbial Consortium with Autofluorescent and Mineralizing Activity for the Biodegradation of Organophosphates, *Journal of Agricultural Food and Chemistry*, Vol.56, No.17, (September 2008), pp.7897–7902, ISSN 0021-8561.

4

Pesticide-Soil Interaction

Rita Földényi[1], Imre Czinkota[2] and László Tolner[2]
[1]Institute of Environmental Sciences, University of Pannonia, Veszprém,
[2]Department for Soil Science and Agricultural Chemistry, Szent István University, Gödöllő
Hungary

1. Introduction

Modern pesticides have to fulfill two requirements: they have to control weeds, pests, diseases, etc. while not to load the environment. These agents should put out their plant protective effect before their degradation. It results in an apparent contradiction because the stable chemicals are able to become environmental contaminants concentrated mostly in the soil. The retention of the pesticides in the soil is important otherwise they can reach the groundwater by leaching. It means that the vertical mobility of these compounds is strongly correlated with their adsorption behavior. If we want to predict the leachability of a pesticide the first step is the investigation of its chemical stability under the given conditions (soil type, pH etc). It is followed by the second step being the study of its adsorption on the soil.

2. The degradation of pesticides in the soil

Pesticide breakdown in the soil is classified as chemical and biological processes influenced by physical factors leading to very complicated pathways. The decomposition is chemical when it takes place in the absence of living organisms. In this case the reaction is activated by thermal, photochemical, radiochemical, electrochemical factors as well as by the interaction with the soil. Photochemical degradation occurs at the soil surface where the energy of sunlight can be absorbed either directly by the pesticide or indirectly by soil components working as photocatalyst. Since atmospheric ozone absorbs solar radiation below 290 nm only chemicals that absorb above this wavelength can be decomposed directly by sunlight. The interaction with the soil means either the catalytic effect of its constituents (e.g. clay minerals) or reactions with its organic matter content (e.g. redox reactions).

In the presence of living organisms biological degradation can take place. In these pathways mainly the enzyme system of microorganisms (bacteria, algae, fungi etc.) works as biocatalyst. Also other species can contribute to the transformation of the parent compound. For example the roots of plants produce fluids which assist the decomposition.

The microorganisms in the soil are either pesticide susceptible or not. In the latter case they will be killed by the compound and no degradation occurs while only a part of the resistant communities can degrade the pesticide (Pierzynski et. al., 1994). The non-biodegradable compounds are persistent and their presence in the soil results in contamination.

The most notorious example for the persistent pesticides is DDT being one of the chlorinated hydrocarbon type insecticides. This compound and their degradation products (DDD, DDE) can be detected in the soil and sediment even after about 30 year of the contamination (Skibniewska et.al., 2003).

As the soil is contaminated in highest amounts by herbicides present study is focused on the fate of different types of these agents in the soil.

The degradation of pesticides takes place in most cases either on (by oxygen, light) or near to the soil surface (by oxygen, microbial activity). For this reason the residence time of the agent in this active layer is a determinant factor. The conditions in deeper active layer are considered mainly anoxic. It is represented by the laboratory experiments (aeration just during the sampling, 25 °C, and natural light in the lab) in the following sections which are numbered according to the type of the herbicide investigated.

2.1 Chloroacetanilide type herbicides (acetochlor and propisochlor)

The chloroacetanilide type herbicides acetochlor and propisochlor (Fig. 1) are widely used in the world. This is the same in Hungary where they were produced in high amounts until the beginning of the XXI. century. In the literature more information can be found about the decomposition of the active ingredient 1 than about 2 which is an original Hungarian product.

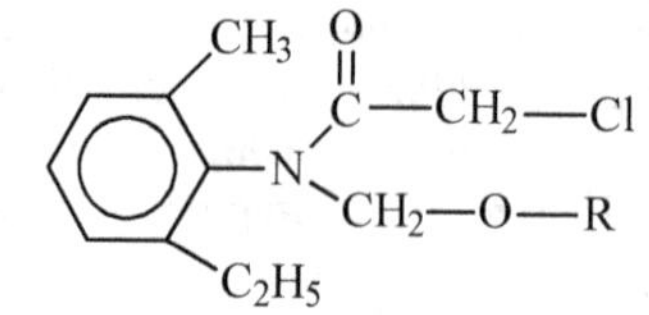

1. R=ethyl
2. R=isopropyl

Fig. 1. The structure of acetochlor (1) and propisochlor (2)

Chloroacetanilides are stable at environmentally relevant pH and temperature in buffer solutions. In the soil, plants and animals more than 30 degradation products of these herbicides have been already identified. Two different dechlorination reactions of acetochlor have been proved in the soil: a. hydrolysis, b. glutathione conjugation (Roberts, 1998a). Both pathways are very important in the detoxification of this compound.

The aim of the study presented here was to check the stability of the compounds 1 and 2 in different types of soils under the conditions detailed above. The decay of these herbicides was followed in the liquid phase (pH=7) of the suspension made from soil : buffer= 1 : 10 ratio. Figure 2 shows that both compounds proved to be rather stable but the degradation of acetochlor (Fig. 2 a.) seemed to be slightly slower than that of propisochlor (Fig. 2 b.). Comparing the soils the decomposition was the fastest in the presence of chernozem having the highest organic carbon content (25.18 mg TOC/g). In the case of propisochlor the order of the breakdown rates follows the TOC values of the soils, which can be seen on Fig 2b. The decay of acetochlor proved to be rather similar in the soils having lower organic carbon content (brown forest and sandy soils).

These results emphasize that the degradation of the chloroacetanilide type compounds need the presence of soil organic matter which is the living space of different species. It means that the degradation is governed by biological processes.

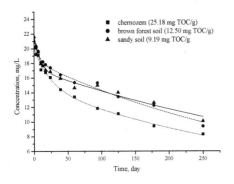 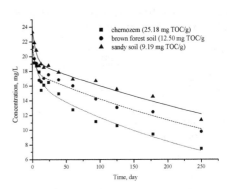

a. Acetochlor degradation, pH=7   b. Propisochlor degradation, pH=7

Fig. 2. Degradation of acetochlor and propisochlor in the presence of different soils

Field measurements in China soils resulted in $t_{1/2} \approx 5$ days for propisochlor (Wang et. al., 2007). In experiments carried out with compound **1** and **2** in soils of temperate zone relatively long half-life time was determined under field conditions: for acetochlor approx. 17 days while for propisochlor approx. 10 days (Ferenczi, 1998). According to these results it can be pointed out that the climate (temperature, moisture, pH of the soil etc.) significantly influences the degradation of these compounds.

We must mention that first order kinetic is commonly used for pesticide decomposition even in the case when the fitting is not the best at the well-known linearization method.

$$c = c_1 \cdot e^{-k_1 \cdot t_1} + c_2 \cdot e^{-k_2 \cdot t_2} \tag{1}$$

c_1 and c_2 : initial concentrations;

k_1 and k_2: rate constants;

t_1 and t_2: reaction time in the two parallel steps.

In our case neither the equation of the simple first order nor that of second order kinetic reaction resulted in relatively good fitting (R^2 should be at least ca. 0.9). It suggests a complex degradation pathway that cannot be described by one process alone. The application of two parallel reactions with first order kinetic (1) resulted in quite good R^2 value (Table 1). The half life time ($t_{i(1/2)}$) was calculated in both steps ($i=1,2$) by equation (2) as it is generally calculated in first-order kinetic reactions.

$$t_{i(1/2)} - \frac{\ln 2}{k_i} \tag{2}$$

where k_i means the rate constants in step 1 and 2.

According to the calculated $t_{i\,(1/2)}$ of these two reactions it can be pointed out that one of the reactions (see $t_{1\,(1/2)}$ in Table 1) is much faster than the other (see $t_{2\,(1/2)}$ in Table 1). Regarding the total degradation rate these data are in accordance with the explanations given for Fig. 2., however, the acetochlor decomposition on brown forest soil is faster than it was understand on the basis of Fig. 2 a.

Our results and the field experiments carried out by other authors (Ferenczi, 1998; Konda & Pasztor, 2001; Wang et. al., 2007) indicate that the organic matter content of the soil is essential in the breakdown of chloroacetanilides but the composition of this organic matter and also other factors can influence this process.

Soil	Chernozem		Brown forest		Sandy soil	
Compound	1	2	1	2	1	2
Parameters						
k_1 (1/day)	0.0426	0.04426	0.13133	0.24598	0.11934	0.18978
$t_{1 (1/2)}$ (day)	16.27	15.66	5.28	2.82	5.81	3.65
k_2 (1/day)	0.00238	0.00273	0.00243	0.00225	0.00189	0.00183
$t_{2 (1/2)}$ (day)	291.24	253.90	285.25	308.07	366.74	378.77
R^2	0.9940	0.9704	0.9874	0.9684	0.9848	0.9774

Table 1. Calculated parameters of equation (1) for chloroacetanilide degradation

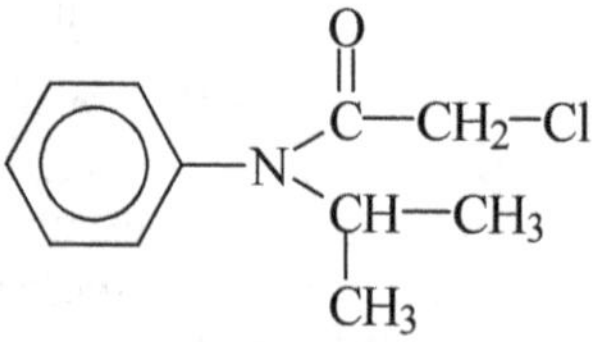

3	4
(2-chloro-[N-(1-methylethyl)-N-phenyl]-acetamide)	(2-hydroxy-[N-(2-ethyl-6-methyl)-phenyl-N-(alkoxy-methyl)]-acetamide)

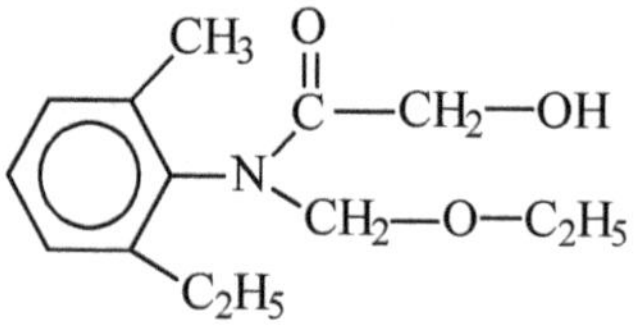

Fig. 3. Chloroacetanilide degradation products identified in the soil

The types of degradation products (Fig. 3) identified in the soil at the end of the experiments support the idea of two ways of pesticide breakdown. (Of course the decomposition pathway can be much more complicated.) After appropriate preparation of the samples compound **3** (m/z: 211, M+) was determined by GC-MS in the case of both chloroacetanilides. The structure of this acetamide indicates an isomerisation reaction. Product **4** determined by DRIFT FT-IR spectroscopy is the result of the hydrolysis of acetochlor.

Isomerization generally needs catalyst that is likely an enzyme of any microorganism in the soil. The hydrolysis can occur in any aquatic medium even under sterile conditions but it can be faster when living organisms are present.

2.2 Urea type herbicides (isoproturon)

The urea type compounds are herbicides with intermediate persistence in aerobic soils. Enhanced rates of their degradation have not been observed even under anaerobic conditions. Isoproturon (see compound **5** in Fig. 5) is applied nowadays most frequently in combination with other active ingredients but even at the end of the XX. century its consume was about 500 tons/year in Europe. Since the information about the decomposition of this compound is rather incomplete (Roberts, 1998b) the present section focuses on this representative of urea type herbicides.

The decay of isoproturon was compared in different buffered solutions (pH=5, pH=7, pH=8) in the absence as well as in the presence of soils. This active ingredient proved to be a rather persistent compound because its concentration in the buffer decreased only by 5.5-12 % until the 350th days of the study. It can be pointed out that the decomposition of isoproturon hardly depends on pH but it is faster in the presence of the soils. Comparing the degradation in the presence of different soils at neutral pH the experience was similar as it was found in the case of chloroacetanilide type herbicides: the organic content of the soil plays leading role in the decay of the compound but also other factors affect the breakdown (Fig. 4, Table 2).

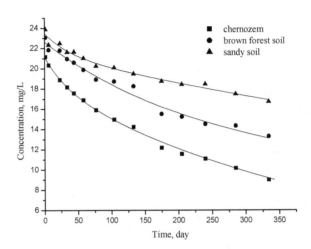

Fig. 4. Degradation of isoproturon in the presence of different soils, pH=7

The fitting according to equation (1) and the determination of the half-life times (Eq. (2)) were carried out as it was described for chloroacetanilide herbicides (Section 2.1.). Calculated parameters show that both steps as well as the total reaction were the slowest in the presence of brown forest soil.

Parameters	Soil		
	Chernozem	Brown forest	Sandy soil
k_1 (1/day)	0.02669	0.0049	0.02517
$t_{1\,(1/2)}$ (day)	25.97	141.46	27.54
k_2 (1/day)	0.00206	0.00057	0.0007
$t_{2\,(1/2)}$ (day)	336.48	1216.05	990.21
R^2	0.9980	0.9817	0.9774

Table 2. Calculated parameters of equation (1) for isoproturon degradation

The investigation of the degradation products, however, indicates other type of reactions, too. According to GC-MS the liquid phase (pH=7) over chernozem contained two degradation products (6 and 7 in Fig. 5). 6 was identified earlier in aquatic solution using UV light (pH>7) as well as in soil while 7 formed only under UV radiation at pH=7 (Roberts, 1998c).

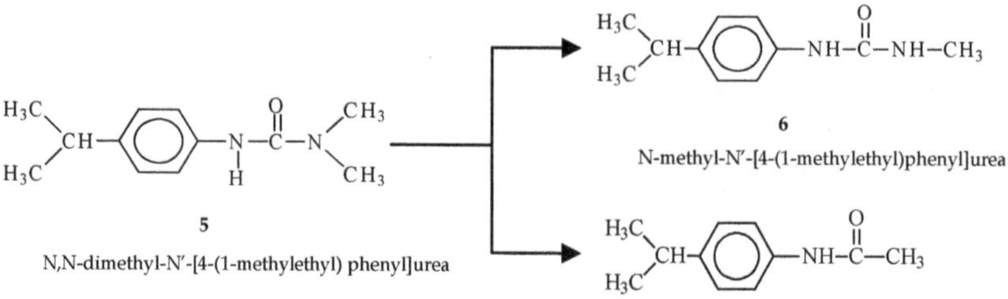

6

N-methyl-N'-[4-(1-methylethyl)phenyl]urea (m/z: 192, M+)

7

N-(4-i-propyl-phenyl)acetamide (m/z: 177, M+)

5

N,N-dimethyl-N'-[4-(1-methylethyl) phenyl]urea

Fig. 5. Isoproturon (5) degradation regarding the products (6, 7) of photo catalytic reaction

This refers to the photosensitization effect of humic substances in the decomposition of isoproturon. This conclusion is in accordance with the results of Gerecke et al. who observed the photocatalytic role of DOM (Gerecke et. al., 2001) in the degradation of phenylurea herbicides. This process can work on the soil surface after application of the herbicide, and in surface waters in case of their pollution by this compound.

2.3 Sulfonyl urea type herbicides

The sulfonyl urea type herbicides (8) are relatively young compounds among the pesticides. Their stability and other chemical as well as physical properties (e.g. solubility) strongly depend on their structure. This is in correlation with the dissociable proton (signed with red color in Fig. 6) that results in weak acidic character of these compounds (Roberts, 1998d).

The importance of this reaction is emphasized by pK_a values. It can be seen e.g. in the comparison of two sulfonyl urea type compounds having just one difference: tribenuron methyl (9 in Fig. 8) has a methyl substitution on N^2 while metsulfuron methyl (10 in Fig. 9) has a proton in this position. The pK_a values of compounds 9 and 10 are 5.0 and 3.3, respectively. For this reason the investigations were focused on the comparison of degradation of these two compounds on two different soils: on the rather basic chernozem (pH=8.34) and on the acidic brown forest (pH=5.92) soils. The aquatic medium equilibrated with the surface was buffered and had the same pH like the soil.

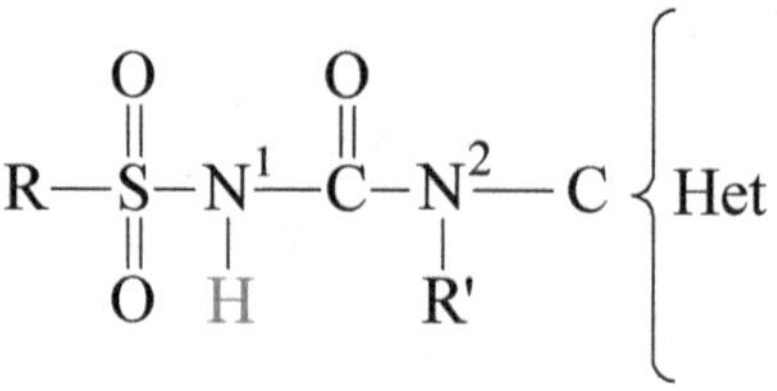

8

R=Me, substituted Ph, substituted heterocycle; C{Het= substituted pyrimidine or triazine, R'=H, Me

Fig. 6. General structure of sulfonylurea herbicides

As it is found the tribenuron methyl is relatively stable over chernozem but not over brown forest soil (Fig. 7). The reaction rates differ much more significantly than in the other cases when the pH was the same and only the TOC content of the soils varied.

The fitting of the measured data for chernozem was carried out according to zero order kinetic being characteristic in heterogeneous system, and it is described with a simple linear equation.

Soil	Chernozem	Brown forest	
Equation	$c=-kt+c_0$	$\ln(c/c_0)=-kt$	$c = c_1 e^{-k_1 t_1} + c_2 e^{-k_2 t_2}$
Parameters	k (1/day) 0.2207	k (1/day) 0.2343	k_1 (1/day) 0.2080
			$t_{1\,(1/2)}$ (day) 3.33
	$t_{1/2}$ (day) 102.70	$t_{1/2}$ (day) 2.96	k_2 (1/day) 3374.74
			$t_{2\,(1/2)}$ (day) 0.0002
R^2	0.9887	0.9934	0.9971

Table 3. Fitting parameters of tribenuron methyl decomposition supposing different kinetic

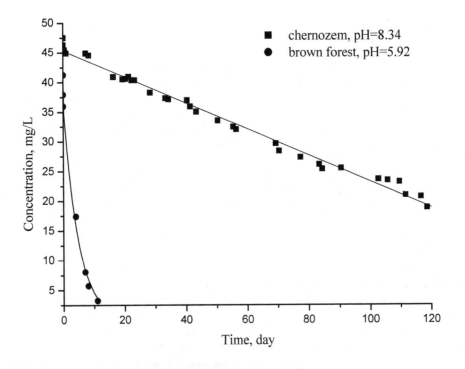

Fig. 7. Degradation of tribenuron methyl in the presence of two different soils

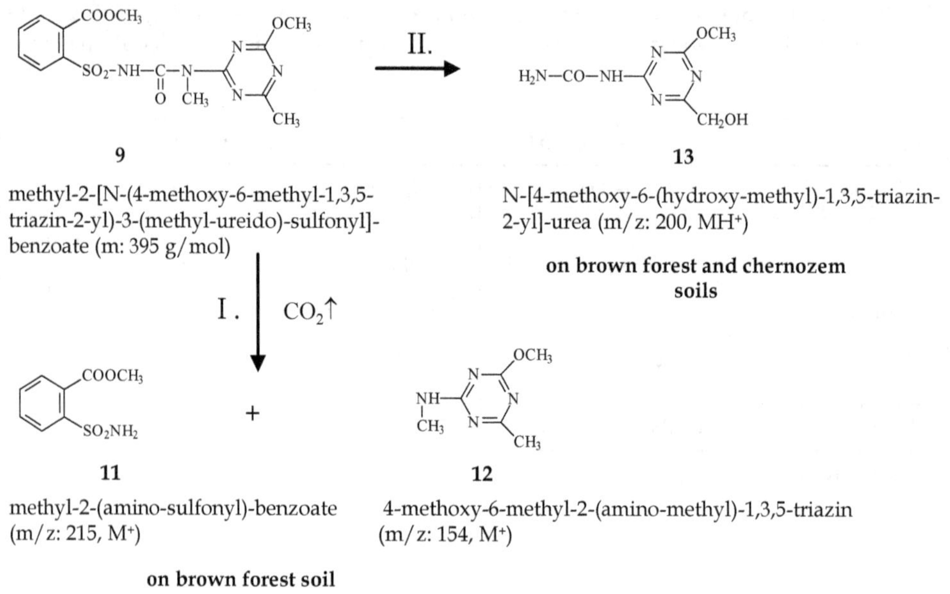

9

methyl-2-[N-(4-methoxy-6-methyl-1,3,5-
triazin-2-yl)-3-(methyl-ureido)-sulfonyl]-
benzoate (m: 395 g/mol)

13

N-[4-methoxy-6-(hydroxy-methyl)-1,3,5-triazin-
2-yl]-urea (m/z: 200, MH$^+$)

**on brown forest and chernozem
soils**

I. $\quad CO_2\uparrow$

11

methyl-2-(amino-sulfonyl)-benzoate
(m/z: 215, M$^+$)

12

4-methoxy-6-methyl-2-(amino-methyl)-1,3,5-triazin
(m/z: 154, M$^+$)

on brown forest soil

Fig. 8. The scheme of tribenuron methyl (**9**) decomposition concerning the identified
degradation products

In the case of brown forest soil the first order kinetic was applied for the fitting when either
one (using linearization method) or two parallel reactions (Eq. (1)) were supposed. Every
counted parameter is summarized in Table 3 where the R^2 values are rather good but
regarding the brown forest soil the fitting for two parallel reactions was even better than
only for one step. According to these results it has to be emphasized that the reaction
pathways are always much more complicated than they are commonly simplified.

The degradation of metsulfuron methyl (**10** in Fig. 9) followed zero order kinetic. This
compound proved to be persistent under the laboratory conditions because its half life
time was over chernozem 500, and over brown forest soil 515 days indicating that the pH
did not influenced its decomposition. Since the decay rate of this compound over the soils
did not show significant difference from that in the buffer solution it seems that the
degradation of this herbicide is slightly assisted by organic matter of the soils used in this
experiment.

The degradation products were investigated after extraction of the equilibrated solution
as well as of the soil and were compared to the blanks studied without soil. According to
GC-MS spectra both extracts (obtained from the solution or from the soil) had the same
components. In the case of the acidic sample (brown forest soil investigations) the
identified degradation products of tribenuron methyl (**9**) were: **11**, **12** and **13** (Fig. 8). The
blank sample contained **11** and **12** indicating the hydrolytic cleavage of the sulfonylurea
linkage (see I. in Fig. 8). It can take place even without the presence of the soil. CO_2 is
produced during this reaction which is often hundreds of times faster under acidic
conditions (Roberts, 1998d).

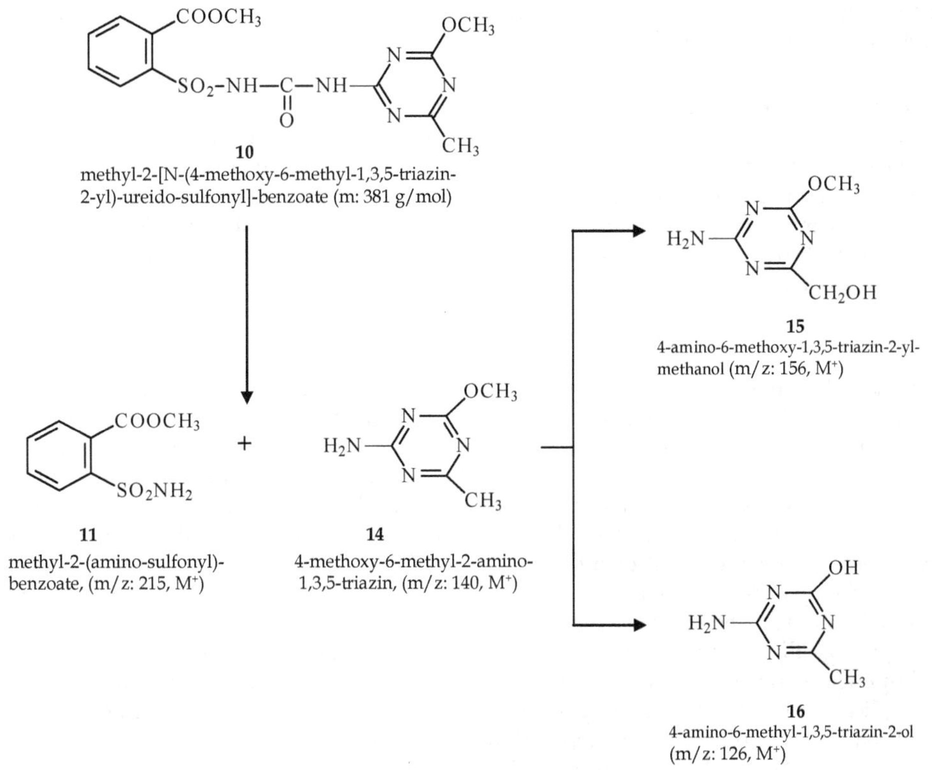

Fig. 9. The scheme of metsulfuron methyl (10) decomposition concerning the identified degradation products

Extracts obtained from the chernozem investigations contained only **13** being a newly determined compound and no hydrolysis products were identified. This result suggests another decomposition mechanism (II.) attributable to the presence of the soil (e.g. humic substances, living organisms). Reaction I. proved in the presence of brown forest soil does not need any organic matter.

Investigations regarding the metsulfuron methyl degradation also resulted in hydrolytic cleavage products of the sulfonylurea linkage (see compounds **11**, **14** in Fig. 9). The other degradation products (**15**, **16** Fig. 9) were found only in the presence of the chernozem soil. These compounds indicate, that the organic matter did not increase the decay rate of metsulfuron methyl, however, it must have a special role in the decomposition.

3. Pesticide sorption on the soil

The sorption on the solid matrix of the soil is one of the most important processes which control transport, persistence, bioavailability and degradation of organic pesticides on soil. Many theories and models have been presented in the literature to describe the different types of sorption isotherms. The most commonly used adsorption isotherm equations for organic contaminants on soil are the Langmuir and the Freundlich isotherms.

The soil as an adsorbent has various active sites leading to rather complicated adsorption mechanisms with the environmental pollutants like pesticides.

3.1 Adsorption described by multi-step isotherm

Bioavailability and environmental transport of pesticides in soil–water system are controlled by their adsorption properties. The extent of adsorption is one of the most important chemical parameter entering into the differential material balance equations of the hydrogeochemical transport models (Weber et al., 1991; Klein et al., 1997; Kovács, 1998). Equilibrium distribution of the studied solute is usually described by some specific form of the general equation (3):

$$q = f(c) \tag{3}$$

where q is the amount of adsorbed solute per gram of sorbent and c is the concentration of the solute in the equilibrium solution phase. Depending on the type of interactions contributing to the adsorption mechanism equation (3) can take on various specific forms. The non-classical nature of the adsorption of hydrophobic organic compounds by soils is taken into consideration in the so-called distributed reactivity model developed by Weber et al. (1992). Soil is basically considered here as a composite material containing inorganic and two types of organic constituents each characterized by its local sorption isotherm. Adsorption of organic contaminants on the exposed surface of the inorganic mineral components is described by the Langmuir isotherm. One part of the soil organic constituent is the geologically older (hard) organic fraction (kerogen, coals etc.) that presents a relatively hydrophobic surface upon which the retention of the contaminant is described by the Freundlich isotherm. The other type of organic soil component is represented by the evolutionary immature (soft) material (humic substances etc.) which is more likely to function as partitioning media upon which the retention of solute is characterized by a linear Henry type of isotherm equation. The distributed reactivity model accommodates these linear and non-linear local adsorption isotherms and the total sorption is approximated as the sum of the isotherms (Weber et al., 1992):

$$q_r = \sum_{i=1}^{m} x_i \cdot q_i \tag{4}$$

where q_r is the overall solid-phase concentration of the solute, q_i is the part of q_r attributable to the ith of m individual local isotherms (in this case it is expressed per unit mass of the solid-phase component responsible for the ith local reaction) and x_i is the mass fraction of that solid-phase component. Different types of isotherms suggested for the description of the adsorption of hydrophobic organic contaminant on soils were summarized and discussed by several recent reviews (Voice & Weber, 1983; Weber et al., 1983; Samiullah, 1990; McBride, 1994; Wolt, 1994; Mader et al., 1997; Carmo et al., 2000). Two-step isotherms were reported for the adsorption of the tenside sodium dodecylsulfate on graphitized carbon (Zettlemoyer & Micale, 1971). At low concentrations the tenside is adsorbed at some equilibrium orientation when a longer length of the hydrocarbon chain is in contact with the surface. At higher concentrations, however, the orientation changes to a new equilibrium position leading to the enhancement of adsorption shown as a second plateau of the isotherm. Systematic study of the adsorption of different surfactants by Clunie & Ingram (1983) and Jönsson et al. (1998a) also revealed the occurrence of two-step isotherms. Sodium dodecyl sulfate (SDS) is a widely used and also well investigated anionic surfactant.

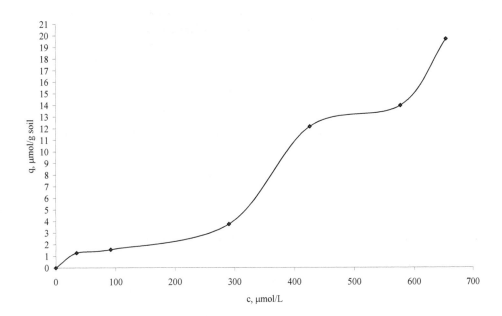

Fig. 10. Adsorption of SDS on sandy soil

Since it can be used even in pesticide formulation the adsorption of SDS was carried out on sandy soil (see Fig. 10). The isotherm has three steps indicating the micelle formation of the molecules either on the surface or in the solution. The critical micelle concentration (cmc) of this surfactant is 830 μmol/L at 20 °C (Jönsson et. al., 1998b). It can be seen on Fig. 10 if the equilibrium concentration is less than 100 μmol/L the curve forms a Langmuir-type isotherm which suggests the filling of active sites by single molecules. When the specific adsorbed amount (q) is 1.6 μmol/g soil the surface is covered by single SDS molecules (maximal coverage: q_T). In the solution of this surfactant the micelle formation starts above 800 μmol/L, however, the molecules adsorbed already on the surface assist the formation of associates even at lower equilibrium concentration values. The second step of the isotherm having 14 μmol/g soil maximal adsorption capacity (q_T) means if the ratio of the q_T values of the second and first steps (14/1.6) is calculated, its result is 9. This allows us to suppose that 9 levels of SDS molecules can cover the surface.

Our experiments with the chloroacetanilide as well as with other type of herbicides usually resulted in similar isotherms to those reported for the adsorption of surfactants. These plots show at least two well defined adsorption steps (see section 3.3), which could not be fitted to the Freundlich or Langmuir equations.

3.2 The derivation of the equation of multi-step isotherm
The aim of the present section is to describe these isotherms by a suitable equation because the steps received little attention in the adsorption studies so far.

In the system reaction (5) is supposed:

$$S + n \times A = SA_n \tag{5}$$

S: empty sites of the soil surface, A: solute, SA_n: surface complex, $n \geq 1$: a non-integer number representing the average degree of association of the solute molecules.
The equilibrium constant (K) of the reaction:

$$K = \frac{[SA_n]}{[S] \cdot [A]^n} \tag{6}$$

where $[S]$, $[A]$, $[SA_n]$ are the equilibrium concentrations of S, A and SA_n, respectively.
The available total concentration of the surface sites $[S]_T$ is given by the mass balance equation (7), and $[S]$ can be substituted:

$$[S]_T = [S] + [SA_n] \tag{7}$$

$$[S] = [S]_T - [SA_n] \tag{8}$$

After the substitution of $[S]$ the value of K can be given by equation (9):

$$K = \frac{[SA_n]}{([S]_T - [SA_n]) \cdot [A]^n} \tag{9}$$

$$[SA_n] = \frac{[S]_T \cdot K \cdot [A]^n}{1 + K \cdot [A]^n} \tag{10}$$

$$q = \frac{q_T \cdot K \cdot [A]^n}{1 + K \cdot [A]^n} \tag{11}$$

where q and q_T are the surface concentrations of $[SA_n]$/mass adsorbent and $[S]_T$/mass adsorbent, respectively.
If associates form above a certain concentration limit of the solute, the concentration variable $[A]$ is replaced by the following relationship:

$$[A] = \frac{c - b + |c - b|}{2} \tag{12}$$

c: controlled value of the equilibrium concentration of solute, b: critical concentration limit of associates.
If $b = 0$ the control variable is given by the concentration of the non-associated solute ($c = [A]$) and associates can be formed only on the surface. If $b > 0$ associates can be formed either on the surface or in the solution and the concentration of these associates appears in equation (13). Its rearrangement results in equation (14) describing one step of the isotherm.

$$q = \frac{q_T \cdot K \cdot \left(\dfrac{c - b + |c - b|}{2}\right)^n}{1 + K \cdot \left(\dfrac{c - b + |c - b|}{2}\right)^n} \tag{13}$$

$$q = \frac{q_T \cdot K \cdot (c - b + |c - b|)^n}{2^n + K \cdot (c - b + |c - b|)^n} \tag{14}$$

Similarly to the so-called distributed reactivity model which allows the addition of various isotherm equations (Weber et al., 1992) the multi-step isotherm is calculated as the sum of the equation of the ith steps:

$$q = \sum_{i=1}^{s} \left\{ \frac{q_{Ti} \cdot K_i \cdot (c - b_i + |c - b_i|)^{n_i}}{2^{n_i} + K_i \cdot (c - b_i + |c - b_i|)^{n_i}} \right\} \tag{15}$$

where q means the specific adsorbed amount, s is the number of steps of the isotherm, q_{Ti} and K_i are the adsorption capacity and the equilibrium constant, while b_i and n_i are the concentration limit and the average degree of association relevant to the ith step of the curve.

3.3 Application of multi-step isotherm equation for the adsorption of certain herbicides on soils and its components

The adsorption properties of those compounds were studied which proved to be stable under conditions being appropriate in static equilibrium experiments. Equation (15) was used to evaluate all of the isotherms. Parameters characteristic for the adsorption curves were calculated by using the nonlinear least square curve fitting procedure of the "Origin" scientific graphing and analysis software.

The adsorption isotherms of chloroacetanilide type herbicides on soils and quartz are shown in Fig. 11. a and b. The two steps of the curves can be clearly seen, and fitting was successful.

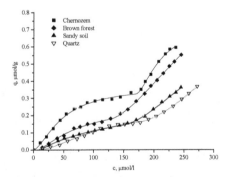 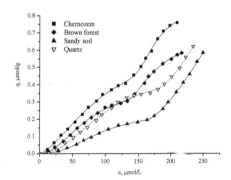

a. Adsorption of acetochlor, pH=7   b. Adsorption of propisochlor, pH=7

Fig. 11. Adsorption of acetochlor (1) and propisochlor (2) on different soils and quartz

The mechanism of the adsorption is very similar in the case of these compounds. More propisochlor is bounded by the soils as well as by quartz than acetochlor. The main force of the adsorption is the hydrophobic interaction.

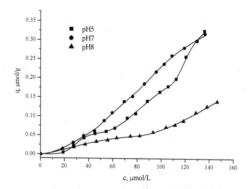

Fig. 12. Adsorption of isoproturon (5) on quartz at different pH values

The adsorption of isoproturon on quartz is introduced here (Fig. 12). The static equilibrium experiments were carried out at different pH values. The points were fitted only once by two- (pH8) and twice by three-step isotherm equation. Since quartz has neutral siloxane surface that functions as a very weak Lewis base (Johnston & Tombácz, 2002) under basic conditions the possibly negatively charged isoproturon is adsorbed in lower amounts than at neutral as well as at acidic pH where the adsorption must be governed by hydrophobic interaction as it was found at chloroacetanilides, too.

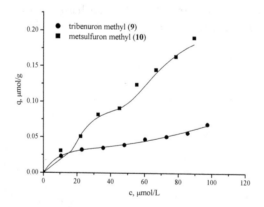

Fig. 13. Adsorption of two sulfonylurea herbicides on chernozem, pH=8.3

Since tribenuron methyl proved to be a rather instable compound under acidic conditions its adsorption studies can give us exact results only at pH ≥ 7. Fig. 13 shows the isotherm of two sulfonylurea herbicides (tribenuron methyl and metsulfuron methyl) on chernozem. The pH of the liquid phase was equal to that of the soil. Metsulfuron methyl (10) has three steps of isotherm and adsorbs in higher amounts on chernozem than the tribenuron methyl (9) having a two-step graph. The structure of 10 results in a more stabile and less polar compound than 9, and the hydrophobic interaction works more effectively leading to more extent adsorption.

4. Buffering of pesticide by soil

The soil has a very important function that is the buffering ability. It means that the soil can adsorb some pesticide as contaminant but it also means if the concentration in the soil solution decreases the pesticides can go to the liquid phase by desorption.

In the literature of soil science and agricultural chemistry the term "buffer capacity" is used mainly for the phosphorus availability and pollution problems. This is the reason that the estimation of the pesticide buffering ability of soils starts here with the equations applied for the better known soil-P system.

The buffer capacity is measured either from the adsorption or from the desorption isotherms and the equilibrium buffer capacity function B is calculable as (16):

$$B = \frac{\partial q}{\partial c} \tag{16}$$

The value B can also be described in the following function of the equilibrium concentration specifically in the case of the Langmuir isotherm (Rattan, 2005):

$$B = \frac{A}{(1 + k \cdot c)} \tag{6}$$

The buffer capacity of the soil-P system can be calculated by means of the phosphorus adsorption isotherm. As a result of the differentiation of the adsorption equation, the equilibrium buffer capacity at any concentration could be calculated as:

$$B = \frac{dP_{ads}}{dc} = \frac{1}{3} \cdot \frac{k}{\sqrt[3]{c^2}} \tag{17}$$

It can be assumed that the adsorption reduces exponentially according to the Freundlich isotherm with $n = 1/3$ exponent.

The advantage of using Q/I relationships (Q: quantity factor – nutrient (P) in the solid phase, I: intensity factor – nutrient (P) in the soil solution) is that they allow the prediction of both P retention and release in soils (Kpomblekou & Tabatabai, 1997). The P-buffering capacity of a soil is its ability to resist a change in the P concentration of the solution phase. Phosphorus-buffering capacities of soils can be related to both plant nutrition and environmental pollution. The Q/I model can be applied to either adsorption or desorption experiments (Yaobing & Michael, 2000). Results showed that Q/I parameters (intercept labile P, a; equilibrium buffering capacity, B; and equilibrium P concentration, EPC) varied significantly between and within sites for the cropping systems studied.

As it is detailed above (see section 3.) the adsorption of herbicides as solutes can not be exactly described neither by simple Langmuir nor by Freundlich equation because these compounds resulted in two- or multi-step isotherms on soils and quartz. Using these isotherms (see equation (15)) the equilibrium buffering capacity (B) was calculated by the derivative function (18):

$$B = \frac{\partial}{\partial c} \left\{ \sum_{i=1}^{s} \left[\frac{q_{T_i} \cdot K_i \cdot \left[(c - b_i) + abs(c - b_i) \right]^{n_i}}{2^{n_i} + K_i \cdot \left[(c - b_i) + abs(c - b_i) \right]^{n_i}} \right] \right\} \tag{18}$$

The model parameters (q_{Ti}, K_i, b_i and n_i) were determined by non-linear regression using sequential simplex optimization.

The mechanical calculation of the derivative of function (18) is impossible because of the break point of abs function. We know, however, that the abs function is just needed due to negative (c-b) data which are not taken into account. It means we can make the derivative function as the sum of single Langmuir isotherms in every x region. Therefore the following function can be applied as the equilibrium buffering capacity of multi-step adsorption isotherms:

$$B = \sum_{i=1}^{s} \left\{ \frac{2^{n_i} \cdot q_{T_i} \cdot K_i \cdot n_i \cdot \left[(c-b_i) + abs(c-b_i) \right]^{n_i-1}}{\left\{ 2^{n_i} + K_i \cdot \left[(c-b_i) + abs(c-b_i) \right]^{n_i} \right\}^2} \right\} \quad (19)$$

Before the interpretation of the calculation of equilibrium buffering capacity of the soil the adsorption isotherms of some pesticides are shown in Figure 14.

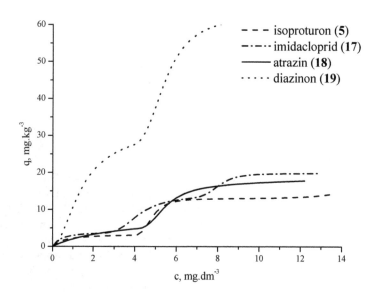

Fig. 14. Adsorption of the investigated pesticides on soil

Sign	q_{T_1}	q_{T_2}	q_{T_3}	K_1	K_2	K_3	b_1	b_2	b_3	n_1	n_2	n_3
5	9.29	9.93	4.42	0.30	3.79	1.47	0	4.57	13.16	0.45	0.55	3.11
17	31.73	31.89	0.00	0.51	0.28	1.00	0	3.86	1.00	1.84	2.67	1.00
18	9.26	10.97	0.00	0.28	0.49	1.00	0	4.16	1.00	0.95	2.46	1.00
19	4.22	9.82	6.01	2.41	0.14	0.01	0	2.33	5.67	1.00	3.00	6.00

Table 4. The fitted parameters of the equilibrium buffering capacity data ($R^2 \geq 0.9960$)

Using the given parameters the equilibrium buffering capacity function was calculated but the larger error of derivative function was fitted on the differentiate data of the original measurement result.

We calculated the new parameters generally (these values are not exactly equal to the original sorption isotherm parameters due to the new fitted weighting points) which are summarized in Table 4.

Calculated functions are shown in Figure 15 where very big differences can be seen in the B values at relatively small equilibrium concentration differences. This phenomenon seems to be almost a periodical change in the case of some compounds.

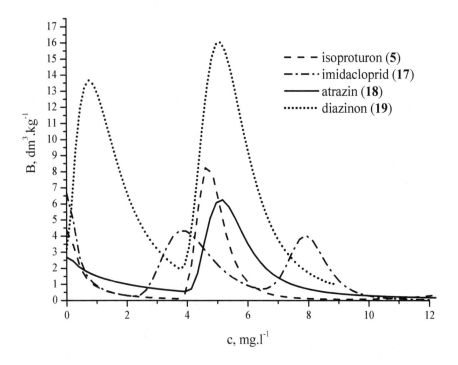

Fig. 15. Equilibrium buffering capacity functions of different pesticides

4.1 An example: The buffering of isoproturon by soil

Adsorption of isoproturon (5) on the soil resulted in three steps of isotherm. In this case the adsorption is described by equation (20) while the buffering capacity by equation (21).

$$q = \frac{\left[q_{T_1} \cdot K_1 \cdot c\right]^{n_1}}{\left[1 + K_1 \cdot c\right]^{n_2}} + \frac{q_{T_2} \cdot K_2 \cdot \left[(c - b_2) + abs(c - b_2)\right]^{n_2}}{2^{n_2} + K_2 \cdot \left[(c - b_2) + abs(c - b_2)\right]^{n_2}} + \frac{q_{T_3} \cdot K_3 \cdot \left[(c - b_3) + abs(c - b_3)\right]^{n_3}}{2^{n_3} + K_3 \cdot \left[(c - b_3) + abs(c - b_3)\right]^{n_3}} \quad (20)$$

$$B = \frac{q_{T_1} \cdot K_1 \cdot n_1 \cdot c^{n_1-1}}{\left\{1+K_1 \cdot [c]^{n_1}\right\}^2} + \frac{2^{n_2} \cdot q_{T_2} \cdot K_2 \cdot n_2 \cdot \left[(c-b_2)+abs(c-b_2)\right]^{n_2-1}}{\left\{2^{n_2}+K_2 \cdot \left[(c-b_2)+abs(c-b_2)\right]^{n_2}\right\}^2} + \frac{2^{n_3} \cdot q_{T_3} \cdot K_3 \cdot n_3 \cdot \left[(c-b_3)+abs(c-b_3)\right]^{n_3-1}}{\left\{2^{n_3}+K_3 \cdot \left[(c-b_3)+abs(c-b_3)\right]^{n_3}\right\}^2}$$

(21)

After substitution of calculated parameters, we can write the equation of the isoproturon adsorption and buffering capacity of examined soil samples as follows (see equation (22) and (23):

$$q = \frac{[9.29 \cdot 0.3 \cdot c]^{0.45}}{[1+0.3 \cdot c]^{0.45}} + \frac{9.93 \cdot 3.79 \cdot \left[(c-3.57)+abs(c-3.57)\right]^{0.55}}{2^{0.55}+3.79 \cdot \left[(c-3.57)+abs(c-3.57)\right]^{0.55}} + \frac{4.42 \cdot 1.47 \cdot \left[(c-13.16)+abs(c-13.16)\right]^{3.11}}{2^{3.11}+1.47 \cdot \left[(c-13.16)+abs(c-13.16)\right]^{3.11}}$$

(22)

$$B = \frac{9.29 \cdot 0.3 \cdot 0.45 \cdot c^{-0.55}}{\left\{1+0.3 \cdot [c]^{0.45_i}\right\}^2} + \frac{2^{0.55} \cdot 9.93 \cdot 3.79 \cdot 0.55 \cdot \left[(c-3.57)+abs(c-3.57)\right]^{-0.45}}{\left\{2^{-0.45} \cdot 3.79 \cdot \left[(c-3.57)+abs(c-3.57)\right]^{0.55}\right\}^2} + \frac{2^{3.11} \cdot 4.42 \cdot 1.47 \cdot 3.11 \cdot \left[(c-13.16)+abs(c-13.16)\right]^{2.11}}{\left\{2^{3.11}+1.47 \cdot \left[(c-13.16)+abs(c-13.16)\right]^{3.11}\right\}^2}$$

(23)

Figure 16 shows the graph of equation (23) emphasizing the special change of B value as a function of the equilibrium concentration.

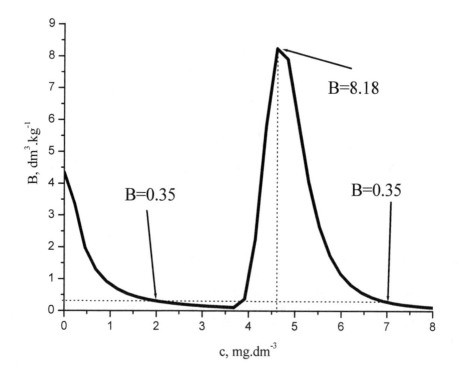

Fig. 16. Equilibrium buffering capacity function of isoproturon

The adsorption isotherm is shown in Figure 17 with some lines for better understanding. The 0.1 size interval of c is marked three times at very important values. The dotted lines show the adsorbed amount intervals connected to the marked equilibrium concentrations. In the B function one maximum and two minimum points can be found (Fig. 16). At the low equilibrium concentration the adsorbed amount of pesticide increased significantly. It means that the slope of the curve (B) is relatively high. Than the surface concentration is increasing and the system is approximating the surface saturation, while the slope (B) is decreasing. The B value reaches its minimum when the surface is almost saturated: in this area the slope of the adsorbed amount function is almost zero (c is approx. 3.5 mg.dm^{-3}). Than the second layer is forming, and the surface concentration will be zero again. New empty sites are present meaning a new type of surface. The adsorbed amount is increasing in function of equilibrium concentration. When the new small adsorbed regions are evolved, the slope of the curve reaches its maximal value: B has its maximum at 4.5 mg.dm^{-1} equilibrium concentration. After this point the process explained above takes place again: the surface is saturating, and the B value is decreasing.

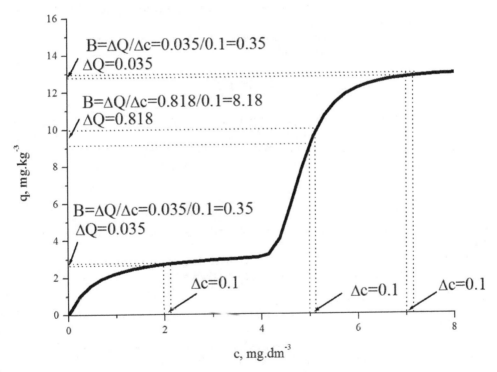

Fig. 17. The explanation of equilibrium buffering capacity function of isoproturon

The parameters and B can be used for calculating of the equilibrium concentration of a contaminated soil while the total amount of a given compound can be calculated from its concentration measured. But this is more complicated, it needs some additional calculations.

If we know the water content of the soil we can give the ratio of adsorbed and dissolved amount of the investigated compound on the basis of the buffering capacity. In the analytical and monitoring practice the change of the concentration in the liquid phase is measured, and it is generally concluded that this is proportional to the changing in the extent of pollution. According to our research this is not true. The suggested calculation method is detailed below step by step.

In this example the amount of the pesticide is present in one unit (1 kg) solid phase, and also one unit (1 dm³) liquid phase. The ratio of solid/liquid phase has to be calculated in the case of known water content of the soil. If the water content Θ is given in dm³.dm⁻³, the volume of solid phase is 1-Θ in dm³.dm⁻³. In this case the following relationship can be written:

$$Q = \frac{1-\Theta}{\Theta} \cdot q \tag{24}$$

where Q is the extent of pollution in solid phase of wet soil, mg.kg⁻¹, q is extent of pollution in clear solid phase of soil (specific adsorbed amount of pollutant), mg.kg⁻¹, Θ is the water content of the soil, dm³.dm⁻³.

The changing of the extent of pollution in solid phase at given water content of the soil can be calculated by the application of equation (24).

$$dQ = \frac{1-\Theta}{\Theta} \cdot B \cdot dc \tag{25}$$

where dQ is he change of the extent of pollution in solid phase, mg.kg⁻¹, dc is the change of the concentration in the solution, mg.dm⁻³, B is the equilibrium buffering capacity, dm³.kg⁻¹. In order to calculate the total amount of the compound in milligrams in the soil we must summarize its amount in the solid as well as in the liquid phase.

$$dQ = \frac{1-\Theta}{\Theta} \cdot V_{soil} \cdot \rho_{soil} \cdot B \cdot dc + \frac{\Theta}{1-\Theta} \cdot V_{soil} \cdot dc \tag{26}$$

where V_{soil} is the total volume of examined soil dm³, ρ_{soil} is density of solid phase of soil, mg.dm⁻³.

In the present example let us assume that we have got one hectare of soil with 25 cm plugged layer, and we calculate total amount of adsorbed material in kilograms.

$$dQ = 10^{-6} \cdot \left(\frac{1-\Theta}{\Theta} \cdot \rho_{soil} \cdot B + \frac{\Theta}{1-\Theta} \right) \cdot V_{soil} \cdot dc \tag{27}$$

where ρ_{soil} is the density of solid phase, about 2.6 kg.dm⁻³; V_{soil} is the volume of soil, in this case 2 500 000 dm³; dQ is the change of the extent of pollution in the given soil area, kg.

Substituting the derived function into B we can calculate the changing buffering effect of the soil in the function of changing equilibrium concentration of the given pesticide. In order to make an approximate calculation, we do not need to use the total complicated function. How we can decide which part should be used from this sum? We have to compare the equilibrium concentration with b_i parameters and we have to choose that part of sum, where b_i is less than the equilibrium concentration, but b_i+1 is bigger than c.

If the equilibrium concentration is less than 3.5 mg.dm-3 the substitution of the known data into equation (27) results in equation (28):

$$dQ = 10^{-6} \cdot \left(\frac{1-0.2}{0.2} \cdot 2.6 \cdot \frac{9.29 \cdot 0.3 \cdot 0.45 \cdot c^{-0.55}}{\left\{ 1+0.3 \cdot [c]^{0.45} \right\}^2} + \frac{0.2}{1-0.2} \right) \cdot 2500000 \cdot dc \qquad (28)$$

where ρ_{soil} is the density of solid phase, about 2.6 kg.dm-3; V_{soil} is the volume of soil, in this case 2 500 000 dm3; Θ is the water content of the soil, dm3.dm-3, in this case 0.2.
If the value of c is: 13.5 mg.dm-3 > c >3.5 mg.dm-3 , the substitution of the known data into equation (27) results in equation (29):

$$dQ = 10^{-6} \cdot \left(\frac{1-0.2}{0.2} \cdot 2.6 \cdot \frac{2^{0.55} \cdot 9.93 \cdot 3.79 \cdot 0.55 \cdot \left[(c-3.57) + abs(c-3.57) \right]^{-0.45}}{\left\{ 2^{-0.45} \cdot 3.79 \cdot \left[(c-3.57) + abs(c-3.57) \right]^{0.55} \right\}^2} + \frac{0.2}{1-0.2} \right) \cdot 2500000 \cdot dc$$

(29)

The calculations above emphasize how big difference can be calculated depending on the equilibrium concentration. It is the reason for using the whole buffering capacity equation and the soil parameters given in equation 28.
Based on the isoproturon curve (Fig. 17.) it can be seen if the equilibrium solution concentration increased from 2 mg.dm-3 to 2.1 mg.dm-3, B is 0.35 dm3.kg-1. The change in the total isoproturon content is 0.1 kg. If the equilibrium solution concentration is increasing from 4.5 mg.dm-3 to 4.6 mg.dm-3 under the same conditions (water content and isoproturon buffer function), B is 8.18 dm3.kg-1. If c changes from 7 mg.dm-3 to 7.1 mg.dm-3, the value of B is 0.35 dm3.kg-1 again. The change in total isoproturon content is 21 kg.
In opposite point of view, for example let's suppose 20 kg.ha-1 isoproturon added into the soil, presumably for plant protection activity. If the original equilibrium solution concentration was 2 mg.dm-3 it is increasing by 2 mg.dm-3 and results in 100 % arising. If the original equilibrium solution concentration was 4.5 mg.dm-3, the increasing is 0.1 mg.dm-3 which means just 2 %. In the case of 7 mg.dm-3 of original equilibrium solution concentration the increasing is 2 mg.dm-3 (35 % arising).

5. Conclusion

The chloroacetanilide type herbicides acetochlor and propisochlor proved to be rather stable compounds under the given conditions. Degradation can occur due to the presence of soil microorganisms. Tribenuron methyl being a sulfonyl urea type compound is stable only under basic conditions and it is decomposed in acidic soils fast. Three degradation products were identified by GC-MS: two of them are formed through acidic hydrolysis and the third newly investigated compound is attributable to the presence of organic matter content of the soil (e.g. humic substances). The structure of metsulfuron methyl is very similar, however, its breakdown is slow. Isoproturon is a rather persistent compound. Its degradation hardly depends on pH but it is faster in the presence of the soil. Two degradation products were identified which can prove the photo catalytic effect of humic substances.
New equation has been derived by making use of the usual mass balance and equilibrium relationships of the adsorption and by considering the possibility of the formation of

associates of the solute molecules. The characteristic model parameters of each step of the adsorption isotherm were estimated for the studied systems by a non-linear least square regression.

Its applicability is proved by the fitting of the isotherms of herbicides having different structures. The calculated curves fit well to the experimentally obtained multi-step isotherms. The parameters of the model can be used for the characterization of the pesticide–soil interactions. On the basis of the new equation formulas were derived to calculate the equilibrium buffering capacity (B) of soil. These buffer capacity equations show special curves with one or more peaks. The measured and fitted values were recalculated by using real soil parameters. These results indicate that we must redefine our original contamination assessment methods in order to avoid even a magnitude error. Not only the equilibrium solution concentration measurement is very important for assessing the real amount of contaminants but also the equilibrium buffering capacity (B) function has to be used.

6. Acknowledgment

The authors gratefully acknowledge their former students – Tímea Ertli, Zsófia Lengyel and Csaba Érsek – for their valuable work.
This chapter was written by the support of the Hungarian National Development Agency (TECH-09-A4-2009-0133, BDREVAM2 project) as well as of the European Union and the European Social Foundation in the frame of the New Hungary Development Plan (TÁMOP-4.2.1.B-10/2/KONV-2010-0001 projects).

7. References

McBride, M. B. (1994). Environmental Chemistry of Soils, p. 342, Oxford University Press, ISBN 0-19-507011-9, New York, USA

Carmo, A. M.; Hundal, L. S. & Thompson, M. L. (2000). Sorption of hydrophobic organic compounds by soil materials: Application of unit equivalent Freundlich coefficients. *Environ. Sci. Technol.* Vol.34, No.20, pp. 4363-4369

Clunie, J. S.; Ingram, B. T. (1983). Adsorption of nonionic surfactants. In: *Adsorption from Solution at the Solid/Liquid Interface.* Parfitt, G. D.; Rochester, C. H. (Eds.), pp. 105 – 152, ISBN 0125449801, Academic Press, London, UK

Ferenczi, J. (1998). Field studies of the behavior of pesticides in the catchment area of the Lake Balaton (in Hungarian). PhD Theses. Pannon University of Agricultural Sciences, Keszthely, Hungary

Gerecke, A. C.; Canonica, S.; Müller, S. R.; Schärer, M. & Schwarzenbach, R. P. (2001). Quantification of dissolved natural organic matter (DOM) mediated phototransformation of phenylurea herbicides in lakes, *Environ. Sci. Technol.* Vol.35, No.19, pp. 3915-3923

Johnston, C. T.; Tombácz, E (2002). Surface Chemistry of Soil Minerals. Chapter 2 In: *Soil Mineralogy with Environmental Applications*, SSSA Book Ser 7, SSSA, Dixon, J.; Schulze, D. (Eds), pp. 37-67, Soil Science Society of America, ISBN 0-89118-839-8, Madison, USA

Jönsson, B.; Lindman, B.; Holmberg, K. & Kronberg,B. (1998). Surfactants and Polymers in Aqueous Solution, (a) pp. 265-294, (b) p. 37, John Wiley and Sons, ISBN 0471-97422-6, Chichester, UK

Klein, M.; Müller, M.; Dust, M.; Görlitz, G.; Gottesbüren, B.; Hassink, J.; Kloskowski, R.; Kubiak, R.; Resseler, H.; Schäfer, H.; Stein, B. & Vereecken, H. (1997). Validation of the pesticide leaching model PELMO using lysimeter studies performed for registration. *Chemosphere* Vol.35, No.11, pp. 2563-2587

Kpomblekou-A K.; Tabatabai M. A. (1997). Effect of cropping systems on quantity/intensity relationships of soil phosphorus. *Soil Sci.* Vol.162, No.1, pp. 56-68

Konda, L. N. & Pasztor, Z. (2001). Environmental distribution of acetochlor, atrazine, chlorpyrifos, and propisochlor under Weld conditions. *J Agr Food Chem*, Vol.49, No.8, pp. 3859–3863

Kovács, B. (1998). Environmental applications of contaminant transport models (in Hungarian). PhD Theses. University of Miskolc, Miskolc, Hungary.

Mader, B. T.; Uwe-Gross, K. & Eisenreich, S. (1997). Sorption of nonionic, hydrophobic organic chemicals to mineral surfaces. *Environ. Sci. Technol.* Vol.31, No.4, pp. 1079-1086

Pierzynski, G. M.; Sims, J. T. & Vance, G. F. (1994). Soils and Environmental Quality, pp. 187-190, Lewis Publishers, ISBN 0-87371-680-9, Boca Raton, USA

Rattan L. (Ed.), (2005). Encyclopedia of Soil Science, Taylor & Francis Group LLC, ISBN 0-84933-830-1, New York, USA

Roberts, T. (Ed.), (1998). Metabolic Pathways of Agrochemicals, Part 1: Herbicides and Plant Growth Regulators, (a) pp. 183-187, (b) p. 706, (c) pp. 735-739, (d) pp. 451-473, The Royal Society of Chemistry, ISBN 0-85404-494-9, Cambridge, UK

Samiullah, Y. (1990). Prediction of the Environmental Fate of Chemicals. Elsevier Science Publishers Ltd., pp. 110-112, ISBN 1-85166-450-5, Barking, UK

Skibniewska, K. A.; Guziur, J.; Grzybowski, M. & Szarek, J. (2003). DDT in soil of pesticide tomb and surrounding ecosystems, *Proceedings of the 10th Symposium on Analytical and Environmental Problems*, pp. 27-31, ISBN 963 212 867 2, SZAB, Szeged, Hungary, September 29, 2003

Voice, T. C.; Weber Jr., W. J. (1983). Sorption of hydrophobic compounds by sediments, soils and suspended solids – I. *Water Res.* Vol.17, No.10, pp. 1433-1441

Wang, S. L.; Liu F. M.; Jin S. H. & Jiang, S. R. (2007) Dissipation of propisochlor and residue analysis in rice, soil and water under field conditions, *Food Control*, Vol.18, No.6, pp. 731-735

Weber, W. J. Jr.; McGinley, P. M. & Katz, L. E. (1991). Sorption phenomena in subsurface systems: concepts, models and effects on contaminant fate and transport. *Water Res.* Vol.25, No.5, pp. 499-528

Weber, W. J. Jr.; McGinley, P. M. & Katz, L. E. (1992). Distributed reactivity model for sorption by soils and sediments. *Environ. Sci. Technol.* Vol.26, No.10, pp. 1955 – 1962

Weber, W. J. Jr.; Voice, T. C.; Pirbazari, M.; Hunt, G. E. & Ylanoff, D. M. (1983). Sorption of hydrophobic compounds by sediments, soils and suspended solids – II. *Water Res.* Vol.17, No.10, pp. 1443 - 1452

Wolt, J. D. (1994). Soil Solution Chemistry, Applications to Environmental Science and
       Agriculture. John Wiley & Sons, Inc., p.169, ISBN 0-471-58554-8, New York, USA
Yaobing S. & Thompson M. L. (2000). Phosphorus Sorption, Desorption, and Buffering
       Capacity in a Biosolids-Amended Mollisol, *Soil Science Society of America Journal*
       Vol.64, pp. 164-169
Zettlemoyer, A. C.; Micale, F. J. (1971). Solution adsorption thermodynamics for organics on
       surfaces. In: *Organic Compounds in Aquatic Environments*. Faust, S.D., Hunter, J.V.
       (Eds.)., pp. 180-181., Marcel Dekker, Inc., ISBN 0824711882, New York., USA

Changing to Minimal Reliance on Pesticides

Paul A. Horne and Jessica Page
IPM Technologies Pty Ltd
Australia

1. Introduction

The most commonly accepted method of controlling pests in Australian crops and throughout the world is the use of pesticides (eg.Pimental 1997, Pimental *et al.* 1997, Page and Horne 2007). There is a reliance on pesticides and the majority of farmers look to chemical companies to continue to provide the familiar means with which they can deal with the complete range of pests that are of concern. In 1997 Pimental *et al* estimated that US farmers used 400 million kg of pesticides annually and worldwide 25 million tonnes of pesticides were applied. An example of the dependence on pesticides can be seen by the close association between the primary producer organizations and pesticide companies and resellers. For example, in Australia, there is an acknowledged strong association between the vegetable industry, as represented by AusVeg (see http://ausveg.com.au) where chemical industry sponsorship of the organization is seen as desirable by both groups. This is alongside an acknowledged effort to implement IPM in vegetables crops in Australia by both AusVeg and several chemical companies. Similarly the Grains Research and Development Corporation (GRDC) in Australia would like to be seen as promoting IPM but the reality is that pesticide applications are the main controls used.

However, there are many examples worldwide of a desire to reduce reliance on pesticides and one particular example of this is by the adoption of integrated pest management (IPM) strategies. It is an approach that is widely seen as desirable and is promoted by many agencies worldwide, including the United Nations, the World Bank and the Food and Agriculture Organisation (FAO) (Maredia 2003, Olsen *et al.* 2003) and also government agencies in Australia (Williams and Il'ichev 2003). Despite this support the change to using this approach is often slow, even though there are examples of success and proven methods of implementing IPM strategies. So, can we successfully reduce our reliance on pesticides if farmers and their advisors do not adopt the strategies that scientists have developed? Also, have scientists developed strategies that are too narrow in design, by dealing with only a narrow pest spectrum?

In this paper we look at what is required to implement change, specifically regarding pest management practices so that there is less reliance on applications of pesticides. The reasons for wanting to reduce reliance on pesticides include insecticide resistance, destruction of natural enemies and other non-target species, residues in produce, environmental concerns and effects on human health (Perkins and Patterson 1997).

1.1 IPM

Integrated Pest Management (IPM) is the use of all available control measures used in a compatible way, but in our opinion should be based on biological and cultural controls with pesticides used only as support tools. The use of these support tools is decided on the results of monitoring of both pest and beneficial species. The basis of this paper is that adoption of IPM is the best method to reduce reliance on pesticides (Perkins and Patterson 1997). This is supported by many publications over many years, (see Table 1 and Horne et al. 2008). So the change to minimal reliance on pesticides depends largely on the adoption of IPM strategies which are based on biological and cultural controls.

Advantages of IPM	Disadvantages of IPM
Reduced dependence on pesticides	More complex than control by pesticide alone and requires a shift in understanding
Increased safety to farm workers, spray operators and the community	Requires a greater understanding of the interactions between pests and beneficials
A slower development of resistance to pesticides	Requires a greater understanding of the effects of chemicals
Reduced contamination of food and the environment	Increased time and resources
Improved crop biodiversity	Level of damage to the crop may initially increase during transition to an IPM programme, in some horticultural crops

Table 1. Advantages and disadvantages of adopting IPM

There are also very many publications that report the fact that levels of adoption of IPM are very often low and rates of adoption are slow (Bajwa and Kogan 2003, Herbert 1995, McNamara *et al.* 1991, Olsen *et al.* 2003, Sivapragasam 2001, Wearing 1988). Even in horticultural crops where the theory of IPM is well developed, achieving widespread adoption on farms remains a challenge (Page and Horne 2007; Boucher and Durgy 2004).). There is a very simple reason for the low rates of adoption of IPM – and it is that the current methods (pesticide applications) are still effective, are legal, and are familiar. There is only a general desire to do something different when this set of reasons is not true. That is, there is a desire to find a new method of pest control when either the pesticides fail (insecticide resistance), or the pesticides are no longer available (withdrawn or banned). Then the pressure to try something unfamiliar becomes more compelling.

The way in which information about IPM is presented, and indeed what is presented, is something that is contentious amongst IPM workers around the world, and we of course have our own opinion. There is a large amount of information available, including web-sites, CD-ROMs, videos, posters and information sheets that are used to provide farmers with the necessary information. Direct contact with farmers is also an option but even this is not an assured method of adoption.

We have a different view on the presentation of IPM information to most (but certainly not all). We believe that there is a scientific basis to the method of presenting IPM that we use, and also that there is a great distinction between information that increases *awareness* of IPM and that which increases *adoption* of IPM.

1.2 What information is presented? Entomologists presenting advice on individual species, farmers dealing with whole crop

An extremely important aspect of advice given about pest management, and in particular pesticide applications, is how wide the consideration is given to the impact of any application. That is, if any action, including pesticide application, was made with the aim of controlling any particular pest, then what consideration is made as to the impact on the control of other pest species? It is not only the degree of control of the target pest achieved by a pesticide application that may have been applied, but also the effect of that pesticide on the biological control agents of other pests.

For example, a key target pest in brassica crops is *Plutella xylostella* (diamondback moth). It is certainly not the only pest in brassica crops but this has been sometimes forgotten by entomologists focusing on this pest. (Given the importance of this species of pest it is understandable that there would be an over-emphasis on this pest alone). The focus was created because of the insecticide resistance that developed in *P. xylostella* in strategies of pest control that were based on insecticide applications alone. An outcome of the singular focus on diamondback moth has been the presentation to brassica growers of strategies labelled in terms such as an: "IPM Strategy for Diamondback Moth". This obviously does not consider the full range of pests, or actions that brassica growers may need to make to deal with a range of pests including other caterpillars and aphids, or the beneficial species that contribute to the control of all of these pests. What is required is an IPM Strategy for Brassica Crops.

The development of strategies based on pesticides to control *P. xylostella* include insecticide resistant management strategies (IRM) and this has been the focus of some researchers in Australia(eg:[http://www.sardi.sa.gov.au/pestsdiseases/horticulture/horticultural_pests/diamondback_moth/insecticide_resistance_management]. Whether or not these strategies work on control of *P. xylostella* is not the only consideration for a brassica grower. To have a brassica such as broccoli or cauliflower accepted by the market there must be adequate control of a range of pests, not just *P. xylostella*. A broccoli or Brussels sprouts crop free of *P. xylostella* but infested with aphids is still not acceptable in the market. What this means of course is that a brassica grower needs to have a strategy to control all pests, not just some. Research entomologists can mistake the target pest on which they are working as the only pest to be considered.

1.3 Disruption of biological control

The IRM strategies that have been developed in various places are examples of how advice on how to control one pest (for example the major pest) can disrupt control of other pests. This is because pesticides that may kill the major pest will also kill the beneficial species that control other pests. This is the most serious problem with pesticide applications in most cases. However, there are other effects that can and do influence the control of pests other than the major pest.

One example is when some of the newer pesticides such as emmamectin ("Proclaim") or indoxacarb ("Avatar") are sprayed for the control of caterpillar pests. These are much more selective than organophosphate and synthetic pyrethroid insecticides and can be used within IPM strategies but their impact on the beneficial species that control other pests can still be significant. For example, emmamectin will kill wasp parasitoids and also predatory bugs such as *Nabis conformis* and indoxacarb will kill a range of beneficial species but is not particularly residual (www.ipmtechnologies.com.au). So the application of these pesticides may achieve the immediate aim of killing target caterpillar pests, but in the process there will be a loss of the biological control agents that are required for on-going and sustainable control. In addition, the control of other pests such as aphids or mites could easily be disrupted because of the loss of a different set of beneficial species. This does not mean these products must never be used but the impact of using them needs to be understood before they are applied. In particular, what species of beneficials are present and is there the possibility of re-invasion of these species?

A common desire amongst many farmers and agronomists, and even some entomologists, is to produce a list rating pesticides as "safe" or "not safe" in absolute terms, not taking into consideration the effect on different species of beneficials. This simplistic approach can be a major setback for IPM adoption as growers tend to simply shift to using less-broad-spectrum insecticides and think they are using IPM. What is required is detailed information on the effects (acute and sub-lethal) of each pesticide on each particular species.

1.4 Awareness versus adoption
There is a tendency amongst some entomologists to concentrate more on awareness of IPM than adoption of IPM. While there needs to be awareness before there can be adoption, the adoption step requires a different set of skills and is something that some organizations promoting IPM would rather avoid. Awareness of IPM may involve talks, workshops, videos, leaflets, manuals, DVD's etc and the production of these tools is a more attractive option to many public organizations as they would seem to be able to reach a large number of people and involve little risk. However, adoption, or implementation, of those practices requires a further step to be taken by the farmer even when there may be awareness of what is possible with IPM. There are different steps needed to implement the change to a totally different method of pest management to that which has been used by the farmer and is familiar.

1.5 What is required to implement change?
In many cases the catalyst for making the change to something unknown (ie IPM) occurs when there is a crisis in pest control, because either the pesticides that have been relied on stop working (insecticide resistance) or the pesticide is no longer available (eg registration withdrawn or the product is withdrawn from sale). While there are other factors that influence the decisions on whether or not to use a pesticide-based strategy (referred to in chapter 2) these two are the most common reasons for a sudden increase in adoption of IPM. Obviously if the methods that are being used no longer work then something different must be done to control pests. In such circumstances farmers are more naturally responsive to embrace a different approach, and examples of different reasons for making changes are given later in this chapter.

It is entirely possible to have farmers understand and implement IPM without any particular crisis. The key factors to success are well documented (eg Herbert 1995) and involves the collaborative and participatory approach to working with individual or small groups of farmers and providing expert, site-specific advice when required.

1.6 Familiar versus different
There is a common experience from a farmer's (or advisor's) point of view when deciding between whether to use a pesticide-based strategy or an IPM strategy. When the starting

point for any individual farmer or farm is that a pesticide-based strategy is known, legal and it works then the decision to adopt something unfamiliar and unproven (on the farmer's own crop) is extremely difficult and is seen as unnecessary and risky. One way to reduce the perceived risk is to have regular monitoring of both pests and beneficial species conducted, in order to check that the desired biological control component of IPM is working as expected.

1.7 The role of monitoring

Monitoring does not control any pests! The reason for saying this is that very often, in our experience, farmers expect that because their crop is being monitored by professional crop monitors then there will automatically be a decrease in pest problems. The only reason for conducting monitoring is to allow timely and informed decisions to be made. Monitoring will allow an assessment to be made about the risk of economic damage by any particular pest. This will include an assessment as to the degree of likely biological control.

1.8 Information on pesticides

It may seem strange at first, but the most important information required to help implement IPM and especially the biological control component, is knowledge about pesticides. This includes information about the impact of pesticides on beneficial species but also involves information about the correct selection of pesticides (including fungicides) for any given problem and in particular to ensure that the pesticide application achieves the result that is desired. Again, it may seem surprising but the practical limitations such as weather and temperature, water volume, adjuvants that should be used with different products are not always (or often) known by farmer or their advisors (including chemical re-sellers).

In an IPM strategy insecticides are used as a support tool when biological and cultural controls are assessed by monitoring as not being sufficient to give adequate levels of control. It is therefore absolutely essential that any application of pesticide is well-timed and correctly applied or else it will not achieve the intended aim of supporting the other control elements. Too often this aspect of IPM is not implemented well. Although research may be done by entomologists (including us) to evaluate the impact of pesticides on beneficial species, it needs to be applied. That is, farmers need to be able to use that information, not just hear about it.

This brings into question the role of entomologists as researchers as opposed to advisors. Worldwide there has been a huge investment in IPM research, and it is the same in Australia. The problem for investors in this research is that there has been a poor level of adoption of IPM as a result of research and the question is why? The answer, we believe, is that researchers do not usually sell or market IPM the way a chemical company or reseller can sell a chemical. What is required is for IPM (and the research that has developed IPM strategies) to be brought into the commercial world rather than stay in the (usually) public funded research environment.

Given that there are only three control options for pest management (biological, cultural or chemical) then it is essential that all three methods work in collaboration, not in opposition. The biological controls in most cases are naturally occurring and cannot be manipulated (apart from not being killed) and cultural controls are usually underestimated. This means that for the farmer, the association between pesticide application and pest control is still paramount. The advisor or crop monitor may know more, but the farmer will be tempted to measure pest control with sprays applied. This means that the decision-making that leads to a pesticide application needs to be absolutely site-specific and take into account the many factors that may be influencing any decision (eg. pest pressure, levels of beneficials, types of beneficials, the observed trend in success of biocontrol, time until harvest, age of crop, time of year, stages of pests, stages of beneficials, intended market etc). These many factors are too many for a general advisory note to include and so site-specific advice is what is needed.

1.9 Collaboration (participatory trials)

We believe that the best way to implement a change to the adoption of IPM is by farmers and advisors working together to obtain site-specific solutions. However, if this is not possible because a lack of such advisors then there are still great changes and advances that farmers can make. With the current technology of the internet, e-mail and digital photos available (and it is certain to improve still more) it is possible to find information and advice relating to the above points. There may not be absolute information regarding particular pest/crop/pesticide/location, but there is usually some information regarding the relative toxicity of pesticides to a range of species (eg. side-effects guides on web-sites such as www.koppert.com). Using the available information a farmer can begin to make changes to what has been the major decision involved in pest management – namely, "what should be sprayed". An option in this decision that involves relying on biological and cultural controls is the option to spray nothing. Information on what pesticides do to a range of beneficial species and not just the pests is the most important tool that most will need to begin to make changes to pest management.

Once the change in pesticide application has been trialled then the farmer may also be more interested in looking at what cultural controls can be utilized along with the different spray regime. Again, information relating to cultural controls can be found or potential (historical/ colloquial) methods trialled on-farm without risking the entire crop.

2. Examples of implementing change: Case studies

2.1 Colin Hurst, Arable cropping farmer, Canterbury, NZ

Colin Hurst runs a 700 ha family farm near Waimate in the southern Canterbury region of New Zealand. The main crops grown are wheat, grass seed, brassica seed and he also grazes sheep. Cropping accounts for 60% of the farm area each year.

Prior to 2006 Colin was operating a conventional system using synthetic pyrethroid sprays ("Karate") as the main defence against aphids and the barley yellow dwarf virus that they can vector. However, in 2006 Colin had an introduction to the concept of IPM via a project initiated by the Foundation for Arable Research (FAR). The initial discussion that aroused Colin's interest was that the sprays applied for aphid control could be causing disruption of the control of other pests, and in particular slugs. In wet years slugs were a significant problem and were expensive to deal with (using baits).

The idea of an IPM approach sounded interesting, especially as it offered an alternative to over-reliance on a single insecticide and the development of resistance, but it was also very different to the mainstream approach being used at the time in New Zealand. The idea of seeing pests increase in number but not applying an insecticide spray was one of the biggest changes to be made and one of the biggest concerns in the early stages. So Colin trialled the approach on half a paddock. This allowed him to see the results of each method and compare the results not only in terms of cost of pesticides (including baits) but also in terms of yield.

The first year's trial was very encouraging, with no pest issues, reduced pesticide use and an increased awareness of beneficial species. Colin had back-up with insect monitoring and decision-making during this time from Plant and Food Research (a government agency in New Zealand) and IPM Technologies P/L visiting from Australia as a part of the project. The realisation that there were many beneficial species present in his crops was something that Colin had not utilised before and he decided that he should investigate further.

Therefore, in the next 2 years of the 3 year project, Colin progressively adopted an IPM approach as he felt more comfortable with the new IPM strategy and decision-making based on monitoring. The monitoring that he uses consists of direct searching for a range of predators and parasitoids, sticky traps to assess what is flying at any time, aphid flight information provided by the Foundation for Arable Research (with Plant and Food Research). Using IPM has meant a much greater use of monitoring than in the past and so any action is based largely on observations on the farm rather than pre- determined sprays or district – wide information. Colin now uses seed dressings of synthetic insecticides rather than sprays as a part of the strategy to use minimal insecticides and instead relies on cultural and biological controls.

The change has been dramatic and in 2010 (4 years after implementing change) the only slug problems requiring treatment was on a border with a neighbours field where there was invasion from outside the farm. The only control measure required in this case was a border application of slug bait. Colin now believes that carabid beetles provide a significant level of control of slugs and is keen not to disrupt this control with sprays targeting aphids.

After 3 years of trials Colin now implements an IPM strategy over the entire farm, and only uses selective insecticides to support the biological and cultural controls as necessary. Although initially daunting, the change in practice has proved worthwhile.

2.2 IPM at Henderson hydroponics, Tasmania

(This is the basis of an article published by Good Fruit and Vegetables magazine in 2009 that describes the change to using IPM with the assistance of the authors of this chapter).

Rob Henderson (and his family) grow hydroponic capsicums near Devonport in Tasmania and experienced major problems during the 2007 – 8 season with tomato spotted wilt virus (TSWV). This virus affects the plant and the fruit, causing affected fruit to be unsalable. The only treatment of infected plants is their removal and disposal which resulted in approximately 75 % of TSWV susceptible cultivars being removed prior to the end of the season, which resulted in considerable financial pain. The problem virus is spread by several species of thrips, including western flower thrips (WFT) which had not previous been present at Henderson Hydroponics. WFT are resistant to many insecticides and that was the problem in this case. The thrips were surviving the insecticides that were used and so were literally out of control. Rob needed to do something different to manage these pests and insecticides did not look like the answer.

In 2008 before his latest crop was planted he met with Dr Paul Horne and Jessica Page of IPM Technologies to discuss implementing an IPM approach. Paul and Jessica were in Tasmania to help develop IPM in a range of vegetable crops and were introduced to Rob by an agronomist (Peta Davies from Roberts Ltd) who saw that this may be the answer to their problem.

A range of predators were introduced throughout the season to control fungus gnats, WFT, aphids and two-spotted mite. It also meant that the broad-spectrum insecticides that he had used in the past could no longer be used, and extreme care had to be taken to ensure that these beneficial species were not disrupted by attempts to control other pests .

Rob admits "We were sceptical at first about IPM, but now we are converts ". He said that "The western flower thrips were present in the latest crop but the predators, in time, controlled them and total damage was reduced to below 2% infection, down from 75% the previous year". Two spotted mite were becoming an increasing problem in previous seasons. However excellent control was achieved with the release of Persimilis, which displayed a ravenous appetite for two spotted mite.

There were some very nervous moments early on in the season when WFT were obviously present and before the predatory mites (known as *cucumeris*) had taken control of the pests. However, the results later in the season speak for themselves and the next seasons expanded crop will again be grown using IPM, but with less nervousness now that Rob knows what to expect.

Rob will be trialling a new thrips predator called Orius that is a new possibility for WFT control. It is being produced in WA.

The project that allowed this to happen is funded by Horticulture Australia and the AusVeg levy.

Rob estimates in his first year of IPM he would have spent approximately treble that which he would have normally spent on insecticide. However he is hopeful that next season, having had a season of IPM experience behind him, this cost may reduce. However as a qualified agricultural economist Rob considers the expenditure on IPM in his greenhouses to be an extremely sound and profitable investment both financially and environmentally.

Henderson Hydroponics staff enjoy working in an environment free of insecticides and have noted the dramatic increase in natural predators, particularly frogs and lady birds seen this season in the greenhouses.

Henderson Hydroponics customers have also been pleased to purchase quality fruit grown without the use of insecticides.

2.3 What has Henderson hydroponics learnt from one season of IPM?

- IPM does work.
- Constant crop monitoring is very important.
- Don't be afraid of seeing small numbers of pests, leave the insecticide locked in the chemical store and only think about using it as a last resort.
- Predators take time to multiply. If pest numbers are increasing, purchase and release more predators rather than waiting for predators to breed up.
- Good advice is readily available. Make use of it. Henderson Hydroponics are extremely grateful for the advice provided by Paul and Jessica during their visits and by them being available to answer questions on the telephone and email at short notice.

3. Conclusion

Reduction of the current reliance on pesticides for the control of pests in agriculture will be best achieved by the adoption of IPM strategies. The poor results (overall) with regard to adoption of IPM is explained in large part by the poor use of known strategies (participatory research) and the fact that IPM is still largely associated with the publicly funded (Government) organisations rather than with commercial aims. It is essential that IPM adoption shifts to the commercial sector to compete with pesticide focussed strategies.

4. Acknowledgements

We thank the many farmers that we have worked with over the last two decades who have allowed us to suggest changes to their pest management practices. We also thank Peter Cole and Neil Hives for discussions on the best methods to implement IPM in a range of crops.

5. References

Bajwa, W.I. and Kogan, M. (2003). IPM adoption by the global community. pp 97 – 107. in *Integrated Pest Management in the Global Arena*. 512pp. Maredia, K.M., Dakouo, D. and Mota-Sanchez, D. (eds). (CABI Publishing, UK.)

Boucher, T.J. and Durgy, R. (2004). Moving towards ecologically based pest management: A case study using perimeter trap cropping. *Journal of Extension* 42, (6) Available online at http://www.joe.org/joe/2004december/a2.shtml

Herbert, D. A. Jr. (1995). Integrated Pest Management Systems: Back to Basics to overcome Adoption Obstacles. *Journal of Agricultural Entomology*. 12, 203 – 210.

Horne, P.A. and Page, J. (2008). *Integrated Pest Management for Crops and Pastures*. 119pp. Landlinks Press, Australia.

Horne, P.A., Page, J. and Nicholson, C. (2008). When will IPM strategies be adopted? An example of development and implementation of IPM strategies in cropping systems. *Australian Journal of Experimental Agriculture*. 48: 1601 – 1607.

Maredia, K. M. (2003). Introduction and overview. pp1-8 in *Integrated Pest Management in the Global Arena*. 512pp. Maredia, K.M., Dakouo, D. and Mota-Sanchez, D. (eds). (CABI Publishing, UK.)

McNamara, K.T., Wetzstein, M.E. and Douse, G.K. (1991). Factors affecting peanut producer adoption of integrated pest management. *Review of Agricultural Economics* 13, 129 – 139.

Olsen, L., Zalom, F. and Adkisson, P. (2003). Integrated Pest Management in the USA. pp249 – 271 in *Integrated Pest Management in the Global Arena*. 512pp. Maredia, K.M., Dakouo, D. and Mota-Sanchez, D. (eds). (CABI Publishing, UK.)

Page, J. and Horne, P.A. (2007). Final Report to Horticulture Australia Limited. Project VG06086: Scoping Study on IPM Potential and Requirements. Available online at http://www.horticulture.com.au

Perkins, J.H. and Patterson, B.R. (1997). Pests, Pesticides and the Environment: A historical perspective on the prospects for pesticide reduction. pp 13 – 33 in *Techniques for Reducing Pesticide Use* 444pp. Pimental, D (ed). (Wiley, UK).

Pimental, D. (1997). Pest Management in Agriculture. pp1 – 11 in *Techniques for Reducing Pesticide Use* 444pp. Pimental, D (ed). (Wiley, UK).

Pimental, D.,Friedman, J. and Kahn, D. (1997). Reducing insecticide, fungicide and herbicide use on vegetables and reducing herbicide use on fruit crops. pp379 – 397 in *Techniques for Reducing Pesticide Use* 444pp. Pimental, D (ed). (Wiley, UK).

Sivapragasam (2001). Brassica IPM adoption: progress and constraints in south-east Asia. Available online at http://www.regional.org.au/au/esa/2001/03/0301siva.htm

Wearing, C.H. (1988). Evaluating the IPM Process. *Annual Review of Entomology* 33, 17 – 38.

Williams, D.G. and Il'ichev, A.L. (2003). Integrated Pest Management in Australia. p371 – 384 in *Integrated Pest Management in the Global Arena*. 512pp. Maredia, K.M., Dakouo, D. and Mota-Sanchez, D. (eds). (CABI Publishing, UK.)

6

The Millardetian Conjunction in the Modern World

Marie-Pierre Rivière[1, 2, 3], Michel Ponchet[1, 2, 3] and Eric Galiana[1, 2, 3]

[1]INRA (Institut National de la Recherche Agronomique), Unité Mixte de Recherche 1301
Interactions Biotiques et Santé Végétale, F-06903 Sophia Antipolis,
[2]CNRS (Centre National de la Recherche Scientifique), Unité Mixte de Recherche 6243
Interactions Biotiques et Santé Végétale, F-06903 Sophia Antipolis,
[3]Université de Nice-Sophia Antipolis, Unité Mixte de Recherche Interactions Biotiques et
Santé Végétale, F-06903 Sophia Antipolis
France

1. Introduction

This chapter deals with the review of literature related to the impact of study of biotic interactions on the development of modern methods to control plant diseases. Only diseases caused by fungi and oomycetes, the two major phylogenetic groups of microbial eukaryotic plant pathogens, were considered. To fight these pathogens, the chemical treatment with fungicides is a long-established method and the most usually used still today. The first report of effective chemical control was related to the use of a fungicide. Its discovery results from the conjunction between a double need, that to protect the vineyards from robbers and downy mildew disease, and a gift for observation which led Millardet to a fertile conclusion for vine protection, but also, for the rise of the chemical treatments of crops. Millardet initially observed that the rows of vineyard, in border of road, treated with an aqueous mixture of copper (II) sulphate to protect against the disease and of lime to dissuade the grape thieves, were preserved from mildew. After experimentations he elaborated a treatment based on a combination of these chemicals now known as the Bordeaux mixture which became the first fungicide (Millardet, 1885; Rapilly, 2001). The mixture is nowadays still used, but it was widely supplanted by synthetic fungicides from various chemical natures (carbamates, triazol, amines, amides, quinines, phenol and benzene derivatives, etc....). Today large quantities of fungicides are applied each year to crops and seeds in the agriculture sector. For example, a mean of 40 000 tons of industrial fungicides are now used each year in France (Aubertot et al., 2005) .

Until the 1980s, the productivist and intensive injunction allowed to nourish the vast majority of the human populations in the developed countries. Because of their low cost and their efficiency, fungicides were used in most countries without restrictions to maximize yield profitably and protected crops. From a phytopathological point of view, plants were mainly looked like simple receptacles, both for the pathogens and the fungicidic molecules. Regarded as a nutritive soup for the first ones and as a simple excipient for the second ones, the protected crop plants laid their fruits with abundance. These last decades, the ecological imperativeness succeeded the productivist one. This has contributed to impose a radically different view of plants in Science and in Agriculture. Host plants are now self-defensing organisms, endowed of an innate immune system, and able to develop various strategies against infections, from the burned ground to the targeted striking. In the same way, substantial knowledge has been gained on the biology of plant pathogens, the epidemiology of diseases and the co-evolution between a host plant and a pathogen. This knowledge constitutes a remarkable sink for genetic and ecological innovations in plant protection. Such alternatives to chemical control have become imperative.

The use of fungicides as well as of the other pesticides (insecticides, herbicides, rodenticides) is now questioned. Their efficiency to control plant pests is counterbalanced by their undesirable and various effects on human health, on sustainability of ecosystems and on biodiversity. There is also the problem of the rapid adaptation of plant pathogenic populations in response to systematic use of pesticide molecules. Within the sustainable development framework, countries and international organizations have a stated political aim of reducing use of pesticides. In France, the Ecophyto 2018 plan constitutes the engagement of the recipients to reduce by 50% the use of the pesticides at the national level within a deadline of ten years, if possible (http://agriculture.gouv.fr/ IMG/pdf/PLAN_ECOPHYTO_2018.pdf, 2008). Several fungicides have already been judged like harmful substances which can cause acute or chronic toxicity. In some cases the marketing authorizations of the preparations containing alarming active substances are withdrawn; their distribution and their use are prohibited. In the European Union the directive 2009/128/EC establishes a framework for community action to achieve the sustainable use of pesticides (http://eur-lex.europa.eu/LexUriServ/LexUriServ.do?uri =CONSLEG:2009L0128:20091125:EN:PDF, 2009). The pesticide program helps government of the Organisation for Economic Co-operation and Development to reduce the risks associated with pesticide use, through a variety of actions to supplement pesticide registration and further reduce the risks that may result even when registered pesticides are properly used (http://www.oecd.org/department/0,3355,en_2649_34383_1_1_1_ 1,00.html). In the same time, the main challenge in agriculture is to increase crop yields for feeding seven billion individuals today and about nine billion on the horizon 2050 (http://km.fao.org/fileadmin/user_upload/fsn/docs/ SUMMARY_2050.pdf, 2009). The impact of the absence of fungicidal protection in plant diseases may reduce crop quality and quantity. The limitation of the fungicide use beyond the optimisation may be harmful for some crops (Butault et al., 2010) (http://www.inra.fr/l_institut/etudes/ ecophyto_r_d/ecophyto_r_d_resultats). The development of integrated pest management, linking all appropriate options including, but not limited to, the judicious use of pesticides, as well as the development of organic food production, limiting the use of pesticides to those that are produced from natural sources, require to prospect new biological resources for plant protection.

The necessity to reduce pesticide use while maintaining high crop yields is today the double need of what we name in this review the millardetian conjunction. What is today the substrate(s), the scientific fields from which could emerge a seminal(s) observation(s) that would supplement the conjunction? Of course it is advisable to say at once how much is hard to anticipate that today. This chapter is focused on the control of plant diseases caused by fungi and oomycetes (Figure 1). Fungi and oomycetes are today mainly and effectively controlled by fungicide applications. We review our knowledge in three domains that we consider as potentially fruitful for such emergence and for rupture in phytoprotection. In the context of studies on plants-pathogens interactions, we underline in the section 2 how this knowledge may help to reduce fungicide use. We also highlight in the section 3 how rapidly expanding investigations on interactions between cells of a pathogen, and between a pathogen and microbial species living in the same biotope may promote environmental friendly innovations in plant protection.

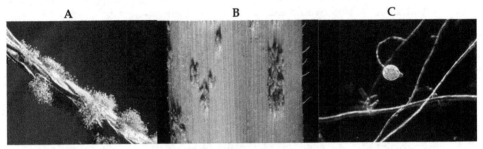

Fig. 1. Examples of eukaryotic pathogens and plant symptoms

(A) Sporulation of the ascomycete *Botrytis cinerea* on a kalanchoe stem. (B) Oat crown rust pustules (basidiomycete)on an oat leaf. (C) Sporangium and mycelium of a polyphagous oomycete, *Phytophthora parasitica*.

2. The plant-pathogen interaction

Beside constitutive physical and chemical barriers preventing infection, plants use their innate responses to ward off pathogens. Plants have evolved the ability to detect microbes through the recognition of conserved microbial leitmotivs which are referred to as Pathogen- or Microbe-Associated Molecular Patterns (PAMPs or MAMPs). The molecular responses are mediated by Pattern-Recognition Receptors (PRRs), a class of innate immune response-expressed proteins that respond to PAMPs. This recognition level initiates MAP kinase signaling and PAMP-triggered immunity (PTI), a key aspect of plant innate immunity which contributes to prevent microbial growth (Nurnberger et al., 2004). Pathogens may suppress PTI responses by secreting effectors in the apoplast or directly into the cytoplasm of host cells, leading to effector-triggered susceptibility (Gohre and Robatzek, 2008). Through evolution and by the driving force of natural selection, plant *R* gene function has emerged resulting in direct or indirect recognition of specific effectors by R proteins. This second level of microbial recognition, specific to certain races or strains of a pathogen, leads to effector-triggered immunity (ETI). ETI is associated in the host with a local programmed cell death, a response which is referred to as the hypersensitive response (HR), and with the establishment in the whole plant of systemic acquired resistance (SAR) which is long lasting and effective against a broad spectrum of pathogens (Chisholm et al., 2006; Dodds and Rathjen, 2010; Jones and Dangl, 2006; Zipfel, 2009).

2.1 Recognition by plants of molecular signatures from pathogens

One way to prevent crop diseases, and in the same time to reduce frequency of chemical treatments, is to enhance the ability of plants to stimulate their own innate immune system. Understanding how plant receptors recognize molecular signatures from pathogens is important to approach such a goal. Over the past 20 years many *R* genes have been discovered and evaluated to engineer disease resistance in crop (Hammond-Kosack and Parker, 2003). On the other hand, only a few plant PRRs have been identified up to now, and our knowledge of the molecular mechanisms underlying PTI is limited. Nevertheless, new agricultural applications could ensue from recent studies on pattern-recognition receptors. A *PRR* gene from the cruciferous plant *Arabidopsis thaliana*, occuring only in the Brassicaceae family, was transferred into two plants, *Nicotiana benthamiana* and *Solanum lycopersicum*, in order to determine if adding new recognition receptors to the host arsenal would lead to better resistance (Lacombe et al., 2010). This *EFR* gene encodes a surface-exposed leucine-rich repeat receptor kinase EFR, and mediates recognition of the bacterial pathogen-associated molecular patterns EF-Tu (elongation factor Tu).   It was chosen by the authors because the high level of conservation of EF-Tu protein sequences across bacteria offered the possibility that EFR could confer resistance against a wide range of bacterial pathogens. Based on triggering of an oxidative burst and on induction of defense-marker genes, expression of EFR in *N. benthamiana* and *S. lycopersicum* transgenic plants was found to confer responsiveness to bacterial elongation factor Tu. The heterologous expression of *EFR* makes also transgenic lines more resistant to a range of phytopathogenic bacteria from different genera (*Pseudomonas, Agrobacterium, Ralstonia, Xanthomonas*). These results were obtained with host plants and pathogens growing in controlled laboratory conditions. Nevertheless, they constitute a first step for the evaluation of the deployment of new PAMP-recognition specificities in crop species. This strategy could be used to engineer pathogen broad-spectrum resistance in crop plants, potentially enabling more durable and sustainable resistance in the field (Dodds and Rathjen, 2010; Gust et al., 2010; McDowell and Stacey, 2008).

2.2 Exogenous application of natural compounds stimulating plant defense responses

As mentioned above, one of the main change in the philosophy of plant disease management has been these last twenty years to abandon the systematic use of biocide treatments against pathogens for alternate solutions among which the bio-activation of plant innate immune system. In some cases it has become reasonable to prevent crop diseases by exogenous application of natural compounds used as elicitors of immune defence responses or of systemic acquired resistance (Vallad and Goodman, 2004). This constitutes a potential alternative or a complement to the intensive use of chemical fungicides with the view to reduce their negative effects on environment and human health. Conventional fungicides are metabolic inhibitors (of electron transport chain, of enzymes, of sterol synthesis of nucleic acid metabolism or protein synthesis) while in contrast elicitors have no direct effect on pathogens. Most of elicitors are natural compounds extracted from microorganisms, algae, and crustacean. Due to their biodegradability and to the low doses applied, the risk of environmental contamination by residues appears weak. Also, they don't show, a priori, a profile to present dangers to human health (Lyon et al., 1995). They appear particularly attractive in the case of integrated production and are evaluated in the frame of the organic farming which lacks anti-fungus substances.

The screening for such natural compounds has led to the characterization of some active molecules now used in the field as a supplement to classic fungicidal treatments. Laminarin, a beta-1→3 glucan, derived from the blue green algae, *Laminaria digitata,* elicits defense responses and resistance to disease in different plants (Aziz et al., 2003; Joubert et al., 1998). Several countries have approved its use particularly on diseases of wheat and barley. Chitosan, another polysaccharide (a deacetylated derivative of chitin, beta-1,4-linked glucosamine) has also been approved by the food and drug administration of the USA first as a wheat seed treatment (El Ghaouth et al., 1994; Hadwiger, 1995). Because of its properties to activate various plant defense responses (phenyl ammonia lyase and peroxidase activities, phytoalexins synthesis, cell wall lignifications) and to trigger resistance, it is considered as an interesting alternative for enhancing natural resistance against *Botrytis cinerea* and other pathogens (Aziz et al., 2006; Povero et al., 2011). Harpin is a proteinaceous stimulator of plant defenses, produced by the plant pathogenic bacterium, *Erwinia amylovora*. When applied to plant surfaces by conventional means, harpin may elicit resistance to pathogens and insects and also enhances plant growth (Wei and Beer, 1996; Wei et al., 1992). Its use is approved in United States on a series of diseases for a wide range of plants : cotton, citrus, wheat, tomatoes, cucumbers, rice, strawberries, peppers, tobacco.

While these elicitors interfere or are suspected to interfere with the early step of recognition by plants of microbial molecular signatures, downstream events of defense signaling pathways have also been subjected to molecular dissection as well as technological evaluation for improving plant resistance to diseases. Two molecular entities have been particularly studied: the *NPR* (for Nonexpressor of *PR* genes) gene family and the salicylic acid (SA), two key positive regulators of systemic acquired resistance (Cao et al., 1994; Vernooij et al., 1994). Salicylic acid has been identified by several lines of evidence as a positive component playing an essential role in the SAR transduction pathway. SA levels are elevated at the onset of SAR in cucumber (Metraux et al., 1990; Rasmussen et al., 1991), tobacco (Malamy et al., 1990), and *Arabidopsis* (Uknes et al., 1993). The exogenous application of SA to leaves of tobacco or *Arabidopsis* induces resistance against the same spectrum of pathogens and activates the same set of SAR genes, as with pathogen-induced SAR (Ward et al., 1991). Transgenic plants expressing a bacterially derived gene that encodes salicylate hydroxylase (*nahG*), an enzyme that converts SA to catechol, are unable to induce SAR (Delaney et al., 1994; Gaffney et al., 1993). The observation that treatment of plants by exogenous SA induces resistance to viral, bacterial and fungal, particularly biotrophic, pathogens has led to application of SA-induced defense responses in plant protection.   A SA derivative, the BTH, benzo(1,2,3)thiadiazole7carbothioic acid Smethyl ester, is mainly used. BTH activates the same set of defense genes and induce similar wide spectrum resistance with lower phytotoxic effect than SA (Gorlach et al., 1996; Lawton et al., 1996). BTH treatment protects against a broad spectrum of pathogens in several fruit, vegetable crops and ornamental plants (Abo-Elyousr et al., 2009; Brisset et al., 2000; Godard et al., 1999; Hukkanen et al., 2007; Iriti et al., 2005; Małolepsza, 2006; Narusaka et al., 1999).

Members of the *NPR* gene family are also key positive regulators of systemic acquired resistance (Cao et al., 1994; Tada et al., 2008). Genetic studies in *Arabidopsis* have demonstrated that *AtNPR1* encodes an ankyrin repeat protein which is involved in SA perception and downstream SAR responses (Cao et al., 1994; Cao et al., 1997; Ryals et al., 1997). Nuclear localization of NPR1 is essential for SA-induced gene expression (Kinkema et al., 2000). Upon pathogen infection accumulation of SA triggers a change in cellular reduction potential, resulting in partial reduction of NPR1 oligomer to monomers, and then in their translocation in the nucleus where they interact with members of the TGA family of basic Leucine zipper transcription factors (Després et al., 2000; Kinkema et al., 2000) that bind to *PR1* promoter elements. NPR1- mediated DNA binding of TGA factors appears to be critical for activation of defense genes (Fan and Dong, 2002; Jupin and Chua, 1996; Lebel et al., 1998; Qin et al., 1994) among which *PR* genes, which encode antimicrobial effectors (Van Loon and Van Strien, 1999). The potential of over-expression of *AtNPR1* from *Arabidopsis thaliana* or of its orthologues in crop species is a current approach for the development of more resistant cultivars. Over-expression of the *AtNPR1* gene in citrus and of the *MpNPR1* gene in apple increases resistance to citrus canker (Zhang et al., 2010) and to fire blight (Malnoy et al., 2007), respectively. In some cases negative impacts of the *NPR1* expression have been observed in transgenic plants. In apple, the overexpression of *Malus NPR1* does not create detrimental morphological changes, but side effects of overexpression of *NH1* (rice homolog of *AtNPR1*) have been noted in rice. The *NH1* overexpression leads both to constitutive activation of defense genes and developmentally controlled lesion-mimic phenotype (Chern et al., 2005; Fitzgerald et al., 2004). On the other hand, overexpression of *AtNPR1* in *Arabidopsis* not only potentiates resistance to different pathogens, but also enhances plant response to BTH and effectiveness of three Oomycete fungicides: metalaxyl, fosetyl, and $Cu(OH)_2$ (Friedrich et al., 2001). The authors suggest that a combination of transgenic and chemical approaches may lead to effective and durable disease-control strategies.

Despite their great potential for control of diseases, treatments of crops with elicitors are not however considered as the panacea for replacing fungicide application. It can be rather considered as a fungicide supplement when fungicide application may be reduced. Indeed treatments with elicitor provide between 20 and 85% disease control and in several cases their application provides no significant level of resistance. To improve their efficiency in the field, information of the influence of the environment, plant genotype, and crop nutrition on plant responses leading to effective resistance remains required (Walters et al., 2005).

2.3 Disease management and plant developmental resistance
In this section we have paid particular attention on knowledge on plant developmental resistance. An increasing number of studies show that induction of resistance to disease during plant development is widespread in the plant kingdom (see for review Develey-Riviere and Galiana, 2007; Panter and Jones, 2002; Whalen, 2005). The scientific community that has investigated this question has used enough diversified approaches, from genetics to epidemiology, to delineate possible and robust contributions of this field for reducing fungicide uses in crop protection.

2.3.1 A parameter for modeling epidemics and to minimize chemical use
One important exciting and difficult challenge in plant protection is to define epidemiologic state both to ensure high crop yields and to manage chemical treatments. A precise definition of the defense and resistance potential of each host plant throughout its life cycle is a key element for the control of pathogen infection. In the context of the ecological awareness, developmental resistance may be considered as a very important factor in the rationalization of cultural practices, the main statement being to reduce fungicide application to shorter periods of high host susceptibility. To achieve this, at least two time parameters have to be properly defined: the precise time point at which establishment of developmental resistance occurs and the length of time during which resistance is effective against the disease. Thus the time required for a plant or for new leaves to acquire developmental resistance is now often integrated as one of variables used in modeling plant diseases (Ficke et al., 2002; Gadoury et al., 2003; Kennelly et al., 2005). For example modeling

of the dynamics of infection caused by sexual and asexual spores during *Plasmopara viticola* epidemics considers that only young grape leaves are receptive to infection because of developmental resistance (Burie et al., 2010; Rossi et al., 2009). Such considerations are also explored for powdery mildew of strawberries. Young leaves, flowers and immature green fruits are much more susceptible to the powdery mildew, caused by the biotrophic fungus *Podosphaera aphanis*, than mature tissues. The high susceptibility to powdery mildew at the early developmental stages seems coincident with the succulent nature of the fruits at this stage, making it easy for penetration and establishment of mildew (Asalf et al., 2009; Carisse and Bouchard, 2010). Control measures targeting at these critical windows of fruit susceptibility are likely to reduce yield loss. The authors of these studies concluded that timing fungicide sprays based on periods of high leaf and berry susceptibility should greatly improve management of strawberry powdery mildew. These few examples illustrate how studies on developmental resistance may help for the development of decision-making tools to minimize environmental and public health risk of fungicide application while maintaining high crop yields.

2.3.2 Genetic tools for breeders

The excavation of various and new genetic resources constitutes an additional window opened by studies on plant developmental resistance. This form of resistance has been now reported for a large number of crop plants. An increasing number of studies have shown that disease resistance governed by major genes (*R* genes) or minor genes (quantitative trait loci, QTLs) may be plant stage-specific. When it occurs the persistence of the phenomenon throughout the rest of the plant life cycle once it has been induced is of clear agronomic interest. The influence of development on race-specific resistance genes has been first studied in detail in rice and wheat, to assist breeders in their decision-making processes (for review Develey-Riviere and Galiana, 2007). A recent finding indicates that QTLs controlling constitutive expression of defense-related genes co-localizes with QTLs for partial resistance of rice to *Magnaporthe oryzae* (Vergne et al., 2010). Such studies also concern other crop plants. Fruits from several cucurbit crops were tested for the effect of fruit development on susceptibility to the oomycete *Phytophthora capsici*. The seven crops tested represent four species: melon (*Cucumis melo*), butternut squash (*Cucurbita moschata*), watermelon (*Citrullus lanatus*), and zucchini, yellow summer squash, acorn squash, and pumpkin (*Cucurbita pepo*). For all of these fruits, a pronounced reduction in susceptibility accompanied the transition from the waxy green to green stage (Ando et al., 2009). The importance to consider developmental resistance for breeding has been underlined in a review on genetic approaches to the management of blister rust (*Cronartium ribicola*) in white pines. The authors have defined developmental resistance, *R*-gene resistance and partial resistance as the three broad categories of resistance that breeders have to take into account for resistance in North American white pines (King et al., 2010).

2.3.3 A putative source for bio-fungicides

Researches on developmental resistance also provided opportunities for characterizing new host molecules influencing pathogen growth *in planta*. Metabolite compounds accumulating in late phases of host plant development may enable the plant to inhibit the infectious cycles of pathogens (Hugot et al., 1999; Kus et al., 2002). However the nature of these compounds remains unknown and it is difficult to define their interest as adaptive resources for plant protection and for their application to crop fields. It has been merely observed that in *Arabidopsis* the intercellular accumulation of SA is critical for antibacterial activity associated with developmental resistance to *Pseudomonas syringae* (Cameron and Zaton, 2004).

2.4 The pathogen in interaction with its host

During the current decade the main research effort on eukaryotic plant pathogens has been and still is the release of genome sequences for pathogens causing the most devastating crop diseases (Dodds, 2010). As a result, an increasing number of gene collections involved in regulation of the interaction with host plants as putative PAMP or effectors have been identified. The identification within these collections of effectors that are crucial for virulence offers the opportunity to select plant targets for more durable resistance (Houterman et al., 2008). In a functional genomics studies Vleeshouvers and coll. (2008) developed an effector-based method for identification of late blight resistance gene in potato. They used a repertoire of secreted and translocated effectors. The putative effectors were predicted computationally from the oomycete *Phytophthora infestans* genome for the presence of a signal peptide and of a RXLR translocation motif into plant cell (Birch et al., 2006; Kamoun, 2006). In an initial set of 54 candidates, two variants of the effector *ipiO*, *ipiO1* and *ipiO2*, were found to trigger HR-associated responses in *Solanum bulbocastanum*, a species carrying the late blight resistance gene *Rpi-blb1*. Both effectors were also found to induce HR responses in *Solanum stoloniferum*, which is the source of the *Rpi-blb1* homologs *Rpi-sto1* and *Rpi-pta1*. The resistance to *P. infestans* cosegregated with response to IpiO in *S. stoloniferum*, and *IpiO* was found to be the avirulence gene of the *Rpi-blb1* resistance gene. Based on these results and on the hypothesis that the resistance genes were orthologous or at least members of the same family, the authors cloned *Rpi-sto1* from *S. stoloniferum* and *Rpi-pta1* from *S. papita* by gene-capture PCR (Polymerase chain reaction). Both genes were found to be functionally equivalent to *Rpi-blb1* and are now used for selective breeding (Pankin et al., 2010).

Comparative genomics of phylogenetically proximal species helps to delineate genome evolution and should also be useful in designing rational strategies for plant disease management. The analytical potential of this approach was illustrated on these two aspects by several articles published in the Science review in 2010 (Dodds, 2010). One of these studies was based on the resequencing of six genomes of four sister species *Phytophthora infestans*, *P. ipomoeae* *P. mirabilis* and *P. phaseoli* (Raffaele et al., 2010). These species infect diverse plants and form a tight clade of pathogens sharing 99.9% identity in their ribosomal DNA internal transcribed spacer region. The aim of the study was to determine how host jumps affect pathogen genome evolution. Genome sequencing allowed the identification of gene-sparse regions and gene-dense regions. Most pathogen genes and genome regions were found highly conserved. But more than 44% of the genes located in the gene-sparse regions showed high diversity suggesting signature of a rapid evolution, when only 14.7% of remaining genes show such signatures. Gene-dense regions were enriched in genes induced in sporangia. Gene-sparse regions were highly enriched in genes induced during plant infection, especially those encoding the predicted RXLR-containing effectors. This is in accordance with the hypothesis that genes induced *in planta* are supposed to evolve faster in a context of a co-evolution with the host. A similar strategy was developed to reveal pathogenicity determinants in two maize smut fungi, *Ustilago maydis* and *Sporisorium reilianum* (Schirawski et al., 2010). These two closely related Basidiomycetes species present an example of differentiation of two closely related pathogens parasitizing the same host. Both genomes were compared and variable genomics regions were identified. These regions were supposed to contain genes encoding virulence proteins since one could expect that pathogen secreted effectors should rapidly evolve. On the other hand, both genomes comprise conserved effector genes as expected for pathogens infecting the same host. Eighty nine percent of the *U. maydis* putative effectors are conserved in *S. reilianum*. This statement could enable to target genomics regions involved in virulence on the same host plant and common to the Basidiomycetes. These studies illustrate how comparative genomics allow identifying the biological functions that are evolutionarily the most stable and that could be targeted to create more durable resistance.

Comparative genomics of more distal pathogenic species within a clade, that of the oomycetes, was also fruitful to define signatures associated with adaptation to a particular trait of life, the obligate biotrophy (Baxter et al., 2010; Spanu et al., 2010). The genome of the obligate biotrophic pathogen *Hyaloperonospora arabidopsidis* was sequenced. The identified gene functions were compared to those of three hemibiotrophic *Phytophthora*, *P. infestans*, *P. sojae* and *P. ramorum* (Baxter et al., 2010). Among a total of 14,543 predicted genes in *H. arabidopsidis*, 6882 had no identifiable orthologs in sequenced *Phytophthora* species. Those genes are potentially involved in biotrophic functions. On the other hand the genome of *H. arabidopsidis* showed a drastic reduction in the number of genes encoding enzymes for assimilation of inorganic nitrogen and sulfur, and proteins associated with zoospore formation and motility. Unsurprisingly, the drastic reduction also concerns genes involved in pathogenicity encoding for degradative enzymes (such as secreted proteinases or cell-wall degrading enzymes), for necrosis and ethylene-inducing (Nep1)-like proteins (NLPs) and for PAMPs. The *H. arabidopsidis* genome also exhibited no more than 134 potential effector proteins with RXLR cell translocation motifs that likely function to suppress host defenses while they have been found to be hundred in the *Phytophthora* genomes (Jiang et al., 2008; Tyler et al., 2006; Whisson et al., 2007). Only 36% of them showed significant similarity percentages with *Phytophthora* effectors. With the aim of obtaining specific targets of biotrophic oomycetes these genes could represent good candidates.

3. Ex planta biotic interactions and plant health

In Phytopathology the plant-pathogen interaction has caught for a long period the attention of most studies at the molecular level. The aim for controlling disease was to develop the scientific bases for genetic engineering of crops (breeding, genetically modified plants). However at least two other kinds of interactions occur at the host plant surface and are crucial for the disease outcome and also for the development of alternative crop protection strategies. Still today, too few studies deal with these two biotic interactions: (i) the cell-cell interaction governing, within the pathogenic species, the biology of the microorganism; (ii) the diverse interactions between the pathogen and the microbial community in their shared habitat. In support to this observation we investigated the features of literature on biotic interactions and plant disease outcome based on bibliometric means. The MEDLINE database was searched via the PubMed access for articles indexed under the publication type "Plant Fungus". Growth of the literature and thematic distribution were addressed. From 1980 to 2010, a total of 35,767 citations were retrieved dealing with a plant-fungus interaction. The literature growth rate is gradually and exponentially growing (Figure 2A). Throughout this period, studies on microbial community and on cell signaling in pathogenic fungi are scarce. These two topics represent respectively 1 % and 1.9% of the whole analyzed literature (Figure 2A and 2B), and for the topic "cell signaling", most of

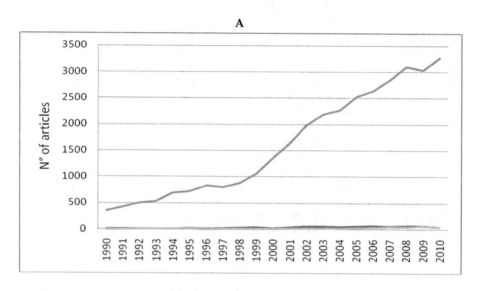

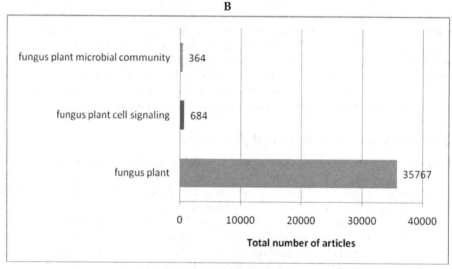

Fig. 2. Bibliometric of the literature on plants-fungi interactions (19. 02. 2011).

PubMed was used to access to the MEDLINE database for searching articles under the publication category "Fungus Plant" (blue), "Fungus Plant Microbial Community" (red), "Fungus Plant Cell Signaling" (red). The retrieved articles were counted and analyzed using Microsoft Excel. (A) growth of the Plant Fungus literature, 1999–2010. (B) Overall number of articles per category. The volume of the literature related to the "Fungus Plant Microbial Community" category is surely underestimated (a search for "Plant Fungus biological control" led to retrieve 1,729 articles but most of them however did not question microbial community as an entity). Similar results were obtained when the data was screened for the publication category "Plant Oomycete" (data not shown).

studies concern the host or fungus responses during the interaction with 46% of them strictly dealing with the plant responses. Thus investigations on interactions between cells of a pathogen and between pathogenic species and the other microorganisms sharing the same biotope are insufficient. Nevertheless, biology and microbial ecology of pathogens must offer opportunities to extend our knowledge on causal relationships between biotic interactions and the epidemiology of a disease. They also open new ways for development of new sustainable agro-ecosystems that should have both agricultural value, by preventing disease, and ecological value, by reducing environmental risks.

3.1 The cell-cell interaction within a pathogenic species
3.1.1 Target cell-to-cell signaling to slow microbial adaptation to treatment
Before infection, cells, spores, zoospores, or mycelia from an eukaryotic pathogen live mainly in groups attached to surfaces, each biological entity interacting with its neighborhood. The influence of these interactions within the species have been until recently neglected, the pathogen being considered at the unicellular level for investigating the interaction with the host. However, prokaryotes and microbial eukaryotes naturally form multicellular aggregates in particular on the surface of putative hosts. The study of these aggregates has shown that microorganisms are capable of complex differentiation and behaviors. The cells communicate and cooperate to perform a wide range of multicellular behaviors, such as dispersal, foraging, biofilm formation, quorum sensing (Atkinson and Williams, 2009; West et al., 2007), these behaviors contributing to the virulence as well as to the dynamics of interactions with the host.

For phytopathogenic bacteria, it has been shown that aggregation promotes virulence in *Ralstonia solanacearum* (Kang et al., 2002), that quorum sensing regulates a variety of virulence factors in *Pectobacterium atrosepticum* (Liu et al., 2008) and that the transition from an aggregated lifestyle to a planktonic lifestyle promotes dissemination in *Xanthomonas campestris* (Dow et al., 2003). As a consequence, the molecular machinery for cell-to-cell signaling constitutes a novel target for the design of antagonists able of attenuating virulence through the blockade of bacterial cell-cell communication (Williams et al., 2000). As mentioned above, cell-to-cell signaling is not limited to the bacterial kingdom. Oomycetes produce and use molecules to monitor population density of biflagellate motile cells, the zoospores. These cells coordinated their communal behaviors by releasing, detecting, and responding to signal molecules (Kong and Hong, 2010). In the species *Phytophthora parasitica*, zoospores may form biofilms on the host surface, using a quorum sensing-like phenomenon to synchronize behavior (Galiana et al., 2008; Theodorakopoulos et al., 2011). Whether such cell-to-cell interactions contribute to the virulence of oomycetes or fungi is not known. However, the fact that, as bacteria, cells of eukaryotic pathogens cooperate to perform multicellular behaviors, indicate that from the dissection of related transduction pathways could emerge new tools for the management of cellular populations on the surface of host plants. Treatments against disease targeted to cell-cell signaling machinery could have an additional benefit and not the least, that to circumscribe the problem of pathogen resistance to fungicides for a larger period, to some extent at least. By performing modeling of multicellular organization in bacteria as a target for drug therapy to predict the speed of resistance evolution, André and Godelle (2005) concluded that this adaptation may be several orders of magnitude slower than in the case of resistance to usual antibiotics. The hypothesis of the authors makes sense in the context of the hierarchical selection theory (Gould, 2002). By targeting treatments against adaptive properties of groups instead of individuals, the relevant unit of organization generating resistance and submitted to selection shifts one level up. Instead of facing billions of cells with a very rapid evolutionary rate, these alternate treatments face a reduced number of larger organisms with lower evolutionary potential (André and Godelle, 2005). Nevertheless, to our knowledge the molecules for such treatments are not yet available for eukaryotic pathogens, and anyway it would be advisable to be sure that they have not pleiotropic or toxic effects.

3.1.2 Biomimetism to trap pathogens

The formation of biofilms is a widely spread property of microbial life governed by cell-cell signaling (Costerton et al., 1999; Danhorn and Fuqua, 2007; Hall-Stoodley et al., 2004; Harding et al., 2009). Biofilm generation is a high spot of research because these structures represent for pathogens an important influence on the virulence as well as on the dynamics of interactions with hosts (Costerton et al., 1999; Hall-Stoodley and Stoodley, 2005). They constitute microbial communities living in co-operative groups attached to surfaces and embedded in a self-producing polymeric matrix. Their formation involves first that planktonic (free-swimming or free-floating) cells become attached to a solid surface, leading to the formation of microcolonies, which then differentiate into exopolysaccharide-encased and fluidfilled channel-separated mature sessile biofilms. Biofilms confer several advantages to pathogens promoting attachment, dissemination or virulence and protecting cells against host defenses and biocide treatments. For human pathologies the failure to eradicate them by standard antimicrobial treatments results in several cases in development of chronic and nosocomial diseases (Costerton et al., 1999; Davies, 2003). The impact of biofilm persistence is not really appreciated for the epidemiology and management of plant diseases (Ramey et al., 2004). To our knowledge nothing is known about potential antimicrobial resistance mechanisms to thwart the efficiency of treatments with fungicides or bactericides. But researches on biofilm may offer an attractive option to diversify biologically-based alternatives to systematic treatments with synthetic fungicides. During the biogenesis of biofilms by an eukaryotic plant pathogen concomitant cellular processes are mobilized to synchronize cell behaviour: chemotaxis, adhesion and aggregation (Galiana et al., 2008; Theodorakopoulos et al., 2011). The elucidation of molecular aspects of these processes should help to elaborate biomimetic materials for the development of trapping systems for pathogens, exactly on the same principles than for the design of insect traps used for many years to monitor or reduce insect populations and based on behavioural confusion techniques (Silverstein, 1981).

3.2 The interactions of the pathogen within a microbial community

In an ecosystem, a plant pathogen evolves within a microbial organized community which has a great influence on the local environment and disease. Before infection various species interact with the pathogen on the host surface shaping the distribution, density and genetic diversity of the inoculum. Such a community is considered and studied as a driving force for natural selection and pathogenicity (Kuramitsu et al., 2007; Siqueira and Rocas, 2009). Concomitantly present metagenomics studies of soils provide pictures of a community structure. The abundance distribution and total diversity can be deciphered. The analyses of the released datasets open a great opportunity to explore into the enormous taxonomic and functional diversity of environmental microbial communities (Simon and Daniel, 2011). By combining studies on function and structure of soil communities, it becomes possible to increase our ability to modify disease states and to question practices of fungicides.

We considered here two levels.

The first one is to re-evaluate the analyses of suppressive soils. Pathogen-suppressive soils have been defined as soils in which the pathogen does not establish or persist, establishes but causes little or no damage, although the pathogen may persist in the soil (Cook and Baker, 1983). Examination of the microbial community compositions in soils possessing various levels of suppressiveness has been referred as a population-based approach (Borneman and Becker, 2007). The strategy leads to establish positive correlation between the population densities of some species and suppressiveness levels, suggesting that they may be involved in the disease suppressive process. The exploration of available metagenomic data will change the dimension of such analyses. As transciptome analysis reveals gene networks for particular cellular functions, Metagenomics may help now to characterize microbial species networks for ecosystemic functions such as pathogen-suppressive properties of soils. This should help to reveal the huge potential of suppressive soils for managing soilborne pathogens. Characterization of the potential may be "easy" when biological nature of the suppression is known as illustrated by studies of soils with known chitinase and antifungal activities (Hjort et al., 2010). Metagenomics may also lead to screen uncultured microorganisms from soil which represent a potentially rich source of useful natural products. During the screening of seven different soil metagenomic libraries for antibacterially active clones, long-chain N-acyltyrosine-producing clones were found in each library. Of the 11 long-chain N-acyl amino acid synthases that were characterized, 10 were unique sequences. The heterologous expression of environmental DNA in easily cultured hosts as *Escherichia coli* has then been used by the authors to illustrate the access to previously inaccessible natural products (Brady et al., 2004).

The second level is more prospective. It consists in screening the functional diversity of microorganisms within communities in which pathogenic species evolve in respect to the disease outcome. In soil as in the other biotopes there is a myriad of microorganisms interacting with each other or with the environment, and performing a wide range of functions (organic decomposition, reduction/oxidation of different forms of elements, nitrogen-fixation...). The set of biotic interactions involving a pathogen constitutes a key factor for the natural population dynamics and emergence of pathogenic clones. In most cases this set remains uncharacterized and one great challenge for improving disease control is to identify in it the biotic interactions which contribute to the negative and also positive control of a pathogenic population. For this aim methods for screening microbial communities to select species associated with a pathogen and impacting the related host disease are missing and must be developed. As a contribution to resolve this problem we have developed a selection method and applied it to a soilborne plant pathogen, *Phytophthora parasitica*, for screening the microbial community from the rhizosphere of the host plant *Nicotiana tabacum*. Two of the selected microorganisms interfered with the oomycete cycle. An ascomycete strongly suppressed the tobacco black shank disease and a ciliate promoted the disease (Galiana et al., unpublished results). In this case the efficiency of the method must be further tested by characterizing other species that affect the tobacco disease. It must also be evaluated for other eukaryotic pathogens before giving food for thought on disease control in two directions. Firstly, the identification of the key suppressive microorganisms will help to diversify material for biological control, a method which have been recommended to replace chemical control methods since it is more economical and environmentally sustainable (Fravel, 2005; Herrera-Estrella and Chet, 1999; Shennan, 2008; Weller et al., 2002). The molecules supporting the suppressive activity of microbial species should be analyzed for their bio-fungicide properties and for their impact on human health and on the rest of the microbial environment. Secondly, the produced information will gradually allow revealing the set of species interacting with a pathogen. In the same time, their abundance in each soil, in each biotope could be easily determined through metagenomic approaches. Thus the combination of both parameters, richness and identity of microbial species affecting a disease cycle, should be an important consideration to define the status of the biotic environment with respect to the occurrence of an epidemic. It could be fruitful to define new decision-making tools that will have to be considered by farmers to decide serenely to restrict fungicide applications or not if required.

4. Conclusion

How protect crops against diseases caused by fungi and oomycetes with both agricultural and ecological value? The treatment of this complex question combines a lot of parameters (crop rotation diversification, crop diversity, rationalization of N-fertiliser application, environment, climate, farmers practices…) mainly treated in the frame of integrated plant disease management. This chapter focuses only on what could emerge from studies on biotic interactions in plant pathology for contributing to the reduction of fungicide use, the development of alternative methods and the selection of crops more tolerant to diseases (Figure 3).

The concern about reduction of fungicides came forward very early from the advent of their use. Based on experimentations, in the lab first and then in the field, Millardet and Gayon (1888) recommended to winegrowers to use a Bordeaux mixture less rich in copper (II) sulphate and lime than in the first formulation. The new mixture was at once more adhesive on leaves than the former, without danger for the vineyard (which did not present any more foliage injuries), and more effective against the mildew. Today a trend to achieve significant fungicide reduction is to diminish frequencies rather than doses. The use of forecasting epidemics systems to assist in the timing of fungicide applications may be one of the appropriate tools. Fungicide treatments would be performed only when necessary. But this may be acceptable by farmers if the risk of an epidemic development of the disease is very low. With the increasing resources on several biotic parameters (timing for establishment of plant developmental resistance, dynamics of clones within a pathogenic population, presence and richness of microbial species affecting a disease cycle) there is an urgent need to associate and integrate the related number of variables to develop more refined and integrated models. They could serve as a starting point to carefully decide in which timing and in which biotic environment the gain from a reduced number of fungicide applications will not alter the potential risk of loss resulting from an incorrect control strategy. Another way of decreasing the frequencies is to combine fungicide treatments with exogenous applications of natural compounds stimulating plant defense responses. Studies on this subject appear scarce and few pieces of information are available on the efficiency of such approach.

Different aspects may also be considered for the development of alternative methods. In the field of genetics, greater possibilities now result from determination of crop plants and pathogens genomes for selecting new varieties of plants capable of resisting to eukaryotic pathogens. Functional and comparative genomics programs have expanded the resource of genes that can be used into crop species, as it has been recently illustrated through the development of new PAMP-recognition specificities in agricultural species or through

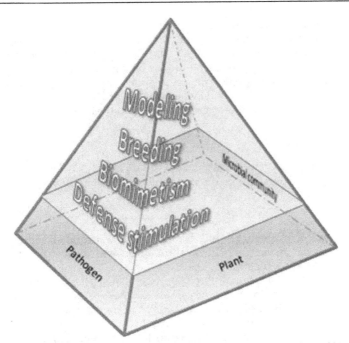

Fig. 3. Schematic representation of the interrelationships between studies of biotic interactions and innovations for crop protection.

The representation by a pyramid symbolizes the integration of different knowledge to develop and properly articulate plant disease management strategies with a low impact on environment and human health. At the bottom the quadrilateral frustum represents the different knowledge of biotic interactions on which may be built crop protection innovations. Three of four base edges mention the biological entities involved in these interactions and which were discussed in this chapter: plant, pathogen, microbial community. The fourth one represents the other biological entities which are important in the biotic environment of a plant (plant community, insects, nematodes,...) and that we did not consider here in the context of the control of disease caused by fungi or oomycetes. At the top of the pyramid are mentioned the topics of emergence of innovations in crop protection. There is no particular consideration for the location of each topic except for modeling at the apex of the pyramid. To our mind this means that robust mathematical models must integrate several biotic variables, often not still parameterized, for building exploitable forecast in terms of rationalization of crop protection.

screening of wild relatives of crop plants to identify new sources of resistance. In the field of biology of organisms, the possibility to elaborate biomimetic materials for the development of behavioural confusion techniques against pathogens must emerge from the molecular elucidation of chemotaxis and aggregation processes. This could lead to design local traps for pathogens associating molecules with specific attractive, aggregative and biocide properties. What is effective to control the populations of insects (pheromone-based trap, sticky fly traps,...) and what was made possible by studies of molecular bases of the behavior of pest insects, must also be effective and possible for the control of pathogens. The validity of disease management by this way should be easily evaluated at the crop scale in hydroponic systems. Hydroponics as an agricultural production system is one of the fastest growing sector, which is more and more used to produce flowers, fruits or vegetable. "Sticky" pathogen traps could contribute to the sanitary quality of the nutrient circulating solutions that is crucial in hydroponic systems. In the field of ecology, new ways could also emerge from exploitation of genomics and metagenomics data to manage pathogenic population in the greenhouse or in the field. Based on appropriate screening of microbial communities they will help to develop and to vary biological control strategies. Beyond the challenge to develop new strategies for crop protection, the biggest defy remains to associate, to articulate them in an adequate way in order to conciliate environmental concerns, safety for human health and agricultural imperativeness.

5. Acknowledgement

This work was supported by a research-aid fund of the CNRS-Cemagref Ecological Engineering program, Programme de recherche interdisciplinaire : « Ingénierie Ecologique».

6. References

Abo-Elyousr, K. A. M., Hashem, M., and Ali, E. H. (2009). Integrated control of cotton root rot disease by mixing fungal biocontrol agents and resistance inducers Original Research Article. Crop Prot *28*, 295-301.

Ando, K., Hammar, S., and Grumet, R. (2009). Age-related resistance of diverse cucurbit fruit to infection by Phytophthora capsici. . J Amer Soc Hort Sci *134*, 176-182.

Andre, J. B., and Godelle, B. (2005). Multicellular organization in bacteria as a target for drug therapy. Ecology Letters *8*, 800-810.

Asalf, B., Stensvand, A., Gadoury, D. M., Seem, R. C., Dobson, A., and Tronsmo, A. M. (2009). Ontogenic resistance to powdery mildew in strawberry fruits. Proceedings of the 10th Annual Epidemiology Workshop - Cornell 4-5.

Atkinson, S., and Williams, P. (2009). Quorum sensing and social networking in the microbial world. J R Soc Interface *6*, 959-978.

Aubertot, J. N., Barbier, J. M., Carpentier, A., Gril, J. J., Guichard, L., Lucas, P., Savary, S., Savini, I., and Voltz, M. é. (2005). Pesticides, agriculture et environnement. Réduire l'utilisation des pesticides et limiter leurs impacts environnementaux. Expertise scientifique collective, synthèse du rapport, INRA et Cemagref (France), 64 p. Available from: <http://www.observatoire-pesticides.gouv.fr/upload/bibliotheque/704624261252893935317453066156/pesticides_synthese_inra_cemagref.pdf>.

Aziz, A., Poinssot, B., Daire, X., Adrian, M., Bézier, A., Lambert, B., Joubert, J. M., and Pugin, A. (2003). Laminarin elicits defense responses in grapevine and induces protection against Botrytis cinerea and Plasmopara viticola. Mol Plant Microbe Interact *16*, 1118-1128.

Aziz, A., Trotel-Aziz, P., Dhuicq, L., Jeandet, P., Couderchet, M., and Vernet, G. (2006). Chitosan oligomers and copper sulfate induce grapevine defense reactions and resistance to gray mold and downy mildew. . Phytopathology *96*, 1188-1194.

Baxter, L., Tripathy, S., Ishaque, N., Boot, N., Cabral, A., Kemen, E., Thines, M., Ah-Fong, A., Anderson, R., Badejoko, W., *et al.* (2010). Signatures of adaptation to obligate biotrophy in the Hyaloperonospora arabidopsidis genome. Science *330*, 1549-1551.

Birch, P. R., Rehmany, A. P., Pritchard, L., Kamoun, S., and Beynon, J. L. (2006). Trafficking arms: Oomycete effectors enter host plant cells. Trends Microbiol *14*, 8-11.

Borneman, J., and Becker, J. O. (2007). Identifying microorganisms involved in specific pathogen suppression in soil. Annu Rev Phytopathol *45*, 153-172.

Brady, S. F., Chao, C. J., and Clardy, J. (2004). Long-chain N-acyltyrosine synthases from environmental DNA. Appl Environ Microbiol 70, 6865-6870.

Brisset, M. N., Cesbron, S., Thomson, S. V., and Paulin, J. P. (2000). AcibenzolarSmethyl induces the accumulation of defenserelated enzymes in apple and protects from fire blight. . Eur J Plant Pathol 106, 529-536.

Burie, J. B., Langlai, M., and Calonnec, A. (2010). Switching from a mechanistic model to a continuous model to study at different scales the effect of vine growth on the dynamic of a powdery mildew epidemic. Annals of Botany in press, 1-11 doi:10.1093/aob/mcq1233.

Butault, J. P., Dedryver, C.-A., Gary, C., Guichard, L., Jacquet, F., Meynard, J.-M., Nicot, P., Pitrat, M., Reau, R., Sauphanor, B., et al. (2010). Synthèse du rapport de l'étude Écophyto R&D Quelles voies pour réduire l'usage des pesticides ? INRA Editeur (France), 90 p. date of acess 28 janvier 2010, Available from: <http://www.inra.fr/1_institut/etudes/ecophyto_r_d/ecophyto_r_d_resultats>.

Cameron, R. K., and Zaton, K. (2004). Intercellular salicylic acid accumulation is important for age-related resistance in Arabidopsis to Pseudomonas syringae. Physiological and Molecular plant Pathology 65, 197-209.

Cao, H., Bowling S.A., Gordon, A. S., and Dong, X. (1994). Characterization of an Arabidopsis mutant that is nonresponsive to inducers of systemic acquired resistance. Plant Cell 1583-1592.

Cao, H., Glazebrook, J., Clarke, J. D., Volko, S., and Dong, X. (1997). The Arabidopsis NPR1 gene that controls systemic acquired resistance encodes a novel protein containing ankyrin repeats. Cell 88, 57-63.

Carisse, O., and Bouchard, J. (2010). Age-related susceptibility of strawberry leaves and berries to infection by Podosphaera aphanis. Crop Protection 29, 969-978.

Chern, M., Fitzgerald, H. A., Canlas, P. E., Navarre, D. A., and Ronald, P. C. (2005). Overexpression of a rice NPR1 homolog leads to constitutive activation of defense response and hypersensitivity to light. . Mol Plant Microbe Interact 18, 511-520.

Chisholm, S. T., Coaker, G., Day, B., and Staskawicz B. J. (2006). Host-Microbe Interactions: Shaping the Evolution of the Plant Immune Response. Cell 124, 803–814.

Cook, R. J., and Baker, K. F. (1983). The Nature and Practice of Biological Control of Plant Pathogens. St. Paul, MN: Am. Phytopathol. Soc. 539 pp.

Costerton, J. W., Stewart, P. S., and Greenberg, E. P. (1999). Bacterial biofilms: a common cause of persistent infections. Science 21, 1318-1322.

Danhorn, T., and Fuqua, C. (2007). Biofilm Formation by Plant-Associated Bacteria. Annu Rev Microbiol 61:, 401-422.

Davies, D. (2003). Understanding biofilm resistance to antibacterial agents. Nature Reviews Drug Discovery 2, 114-122.

Delaney, T. P., Uknes, S., Vernooij, B., Friedrich, L., Weymann, K., Negrotto, D., Gaffney, T., Gut-Rella, M., Kessmann, H., Ward, E., and Ryals, J. (1994). A central role of salicylic Acid in plant disease resistance. Science 266, 1247-1250.

Després, C., DeLong. C., Glaze. S., Liu, E., and Fobert, P. R. (2000). The Arabidopsis NPR1/NIM1 protein enhances the DNA binding activity of a subgroup of the TGA family of bZIP transcription factors. Plant Cell 12.279-290.

Develey-Riviere, M. P., and Galiana, E. (2007). Resistance to pathogens and host developmental stage: a multifaceted relationship within the plant kingdom. New Phytol 175, 405-416.

Dodds, P. N. (2010). Genome evolution in plant pathogens. Science 330, 1486-1487.

Dodds, P. N., and Rathjen, J. P. (2010). Plant immunity: towards an integrated view of plant-pathogen interactions. Nat Rev Genet 11, 539-548.

Dow, J. M., Crossman, L., Findlay, K., He, Y. Q., Feng, J. X., and Tang, J. L. (2003). Biofilm dispersal in Xanthomonas campestris is controlled by cell-cell signaling and is required for full virulence to plants. Proc Natl Acad Sci U S A 100, 10995-11000.

El Ghaouth, A., Arul, J., Grenier, J., Benhamou, N., Asselin, A., and Belanger, R. (1994). Effect of chitosan on cucumber plants: suppression of Pythium aphanidermatum and induction of defence reactions. Phytopathology 84, 313-320.

Fan, W., and Dong, X. (2002). In vivo interaction between NPR1 and transcription factor TGA2 leads to salicylic acid-mediated gene activation in Arabidopsis. Plant Cell 14, 1377-1389.

Ficke, A., Gadoury, D. M., and Seem, R. C. (2002). Ontogenic resistance and plant disease management: A case study of grape powdery mildew. The American Phytopathological Society 92, 671-675.

Fitzgerald, H. A., Chern, M.-S., Navarre, R., and , and Ronald, P. C. (2004). Overexpression of (At)NPR1 in rice leads to a BTH and environment induced lesion mimic/cell death phenotype. Mol Plant-Microbe Interact 17, 140-151.

Fravel, D. R. (2005). Commercialization and implementation of biocontrol. Annu Rev Phytopathol 43, 337-359.

Friedrich, L., Lawton, K., Dietrich, R. , Willits, M., Cade, R., and Ryals, J. (2001). NIM1 overexpression in Arabidopsis potentiates plant disease resistance and results in enhanced effectiveness of fungicides. Mol Plant Microbe Interact 14, 1114-1124.

Gadoury, D. M., Seem, R. C., Ficke, A., and Wilcox, W. F. (2003). Ontogenic resistance to powdery mildew in grape berries. Phytopathology 93, 547-555.

Gaffney, T., Friedrich, L., Vernooij, B., Negrotto, D., Nye, G., Uknes, S., Ward, E., Kessmann, H., and Ryals, J. (1993). Requirement of salicylic Acid for the induction of systemic acquired resistance. Science 261, 754-756.

Galiana, E., Fourré, S., and Engler, G. (2008). Phytophthora parasitica biofilm formation: installation and organization of microcolonies on the surface of a host plant. Environ Microbiol 10, 2164-2171.

Godard, J. F., Ziadi, S., Monot, C., Le Corre, D., and D., S. (1999). Benzothiadiazole (BTH) induces resistance in cauliflower (Brassica oleracea var botrytis) to downy mildew of crucifers (Peronospora parasitica) Crop Prot 18, 397-405.

Gohre, V., and Robatzek, S. (2008). Breaking the barriers: microbial effector molecules subvert plant immunity. Annu Rev Phytopathol 46, 189-215.

Gorlach, J., Volrath, S., Knauf-Beiter, G., Hengy, G., Beckhove, U., Kogel, K. H., Oostendorp, M., Staub, T., Ward, E., Kessmann, H., and Ryals, J. (1996). Benzothiadiazole, a novel class of inducers of systemic acquired resistance, activates gene expression and disease resistance in wheat. Plant Cell 8, 629-643.

Gould, S. J. (2002). The structure of evolutionary theory. Belknap (Harvard University Press), Cambridge, MA) 1401 pp.

Gust, A. A., Brunner, F., and Nurnberger, T. (2010). Biotechnological concepts for improving plant innate immunity. Curr Opin Biotechnol 21, 204-210.

Hadwiger, L. A. (1995). Chitosan as crop regulator. In: MB Zakaria, WMW Muda, MP Abdullah (eds): Chitin and chitosan: the versatile envrinmentally friendly modern materials. Penerbit Universiti Kebangsaan, Malaysia Bangi., 227-236.

Hall-Stoodley, L., Costerton, J. W., and Stoodley. P (2004). Bacterial biofilms: from the natural environment to infectious diseases. Nat Rev Microbiol 2, 95-108.

Hall-Stoodley, L., and Stoodley, P. (2005). Biofilm formation and dispersal and the transmission of human pathogens. Trends Microbiol 13, 7-10.

Hammond-Kosack, K. E., and Parker, J. E. (2003). Deciphering plant-pathogen communication: fresh perspectives for molecular resistance breeding. Curr Opin Biotechnol 14, 177-193.

Harding, M. W., Marques, L. L., Howard, R. J., and Olson, M. E. (2009). Can filamentous fungi form biofilms? Trends Microbiol 17, 475-480.

Herrera-Estrella, A., and Chet, I. (1999). Chitinases in biological control. Exs 87, 171-184.

Hjort, K., Bergström, M., Adesina, M. F., Jansson, J. K., Smalla, K., and Sjöling, S. (2010). Chitinase genes revealed and compared in bacterial isolates, DNA extracts and a metagenomic library from a phytopathogen-suppressive soil. FEMS Microbiol Ecol. 71, 197-207.

Houterman, P. M., Cornelissen, B. J., and Rep, M. (2008). Suppression of plant resistance gene-based immunity by a fungal effector. PLoS Pathog 4, e1000061.

Hugot, K., Aime, S., Conrod, S., Poupet, A., and Galiana, E. (1999). Developmental regulated mechanisms affect the ability of a fungal pathogen to infect and colonize tobacco leaves. Plant J 20, 163-170.

Hukkanen, A. T., Kokko, H. I., Buchala, A. J., McDougall, G. J., Stewart, D., Kärenlampi, S. O., and Karjalainen, R. O. (2007). Benzothiadiazole induces the accumulation of phenolics and improves resistance to powdery mildew in strawberries. J. Agric Food Chem 55, 1862-1870.

Iriti, M., Rossoni, M., Borgo, M., Ferrara, L., and Faoro, F. (2005). Induction of resistance to gray mold with benzothiadiazole modifies amino acid profile and increases proanthocyanidins in grape: primary versus secondary metabolism. J Agric Food Chem 53, 9133-9139.

Jiang, R. H., Tripathy, S., Govers, F., and Tyler, B. M. (2008). RXLR effector reservoir in two Phytophthora species is dominated by a single rapidly evolving superfamily with more than 700 members. Proc Natl Acad Sci U S A 105, 4874-4879.

Jones, J. D., and Dangl, J. L. (2006). The plant immune system. Nature 444, 323-329.

Joubert, J., Yvin, Y., Barchietto, T., Seng, J., Plesse, B., Klarzynski, O., Kopp, M., Fritig, B., and Kloareg, B. (1998). A beta1-3 glucan, specific to a marine alga, stimulates plant defence reactions and induces broad range resistance against pathogens. Proc Brighton Conf, Pests & Dis, 441-448.

Jupin, I., and Chua, N. H. (1996). Activation of the CaMV as-1 cis-element by salicylic acid: differential DNA-binding of a factor related to TGA1a. EMBO J 15, 5679-5689.

Kamoun, S. (2006). A catalogue of the effector secretome of plant pathogenic oomycetes. Annu Rev Phytopathol 44, 41-60

Kang, Y., Liu, H., Genin, S., Schell, M. A., and Denny, T. P. (2002). Ralstonia solanacearum requires type 4 pili to adhere to multiple surfaces and for natural transformation and virulence. Mol Microbiol 46, 427-437.

Kennelly, M. M., Gadoury, D. M., Wilcox, W. F., Magarey, P. A., and Seem, R. C. (2005). Seasonal Development of Ontogenic Resistance to Downy Mildew in Grape Berries and Rachises. J Phytopath 95, 1445-1452.

King, J. N., David, A., Noshad, D., and Smith, J. (2010). A review of genetic approaches to the management of blister rust in white pines. Forest Pathology 40, 292-313. doi: 210.1111/j.1439-0329.2010.00659.x.

Kinkema, M., Fan, W., and Dong, X. (2000). Nuclear localization of NPR1 is required for activation of PR gene expression. Plant Cell 12, 2339-2350.

Kong, P., and Hong, C. (2010). Zoospore density-dependent behaviors of Phytophthora nicotianae are autoregulated by extracellular products. Phytopathology 100, 632-637.

Kuramitsu, H. K., He, X., Lux, R., Anderson, M. H., and Shi, W. (2007). Interspecies interactions within oral microbial communities. Microbiol Mol Biol Rev 71, 653-670.

Kus, J. V., Zaton, K., Sarkar, R., and Cameron, R. K. (2002). Age-related resistance in Arabidopsis is a developmentally regulated defense response to Pseudomonas syringae. Plant Cell 14, 479-490.

Lacombe, S., Rougon-Cardoso, A., Sherwood, E., Peeters, N., Dahlbeck, D., van Esse, H. P., Smoker, M., Rallapalli, G., Thomma, B. P., Staskawicz, B., et al. (2010). Interfamily transfer of a plant pattern-recognition receptor confers broad-spectrum bacterial resistance. Nat Biotechnol 28, 365-369.

Lawton, K. A., Friedrich, L., Hunt M., Weymann, K., Delaney, T., Kessmann, H., Staub, T., and Ryals, J. (1996). Benzothiadiazole induces disease resistance in Arabidopsis by activation of the systemic acquired resistance signal transduction pathway. Plant J 10, 71-82.

Lebel, E., Heifetz, P., Thorne, L., Uknes, S., Ryals, J., and Ward, E. (1998). Functional analysis of regulatory sequences controlling PR-1 gene expression in Arabidopsis. . Plant J 16, 223-233.

Liu, H., Coulthurst, S. J., Pritchard, L., Hedley, P. E., Ravensdale, M., Humphris, S., Burr, T., Takle, G., Brurberg, M. B., Birch, P. R., et al. (2008). Quorum sensing coordinates brute force and stealth modes of infection in the plant pathogen Pectobacterium atrosepticum. PLoS Pathog 4, e1000093.

Lyon, G. D., Reglinski, T., and Newton, A. C. (1995). Novel disease control compounds: the potential to "immunize" plants against infection. Plant Pathol 44, 407-427.

Malamy, J., Carr, J. P., Klessig, D. F., and Raskin, I. (1990). Salicylic acid : a likely endogenous signal in the resistance response of tobacco to viral infection. Science 250, 1002-1004.

Malnoy, M., Jin, Q., Borejsza-Wysocka, E. E., He, S. Y., and Aldwinckle, H. S. (2007). Overexpression of the apple MpNPR1 gene confers increased disease resistance in Malus x domestica. Mol Plant Microbe Interact 20, 1568-1580.

Małolepsza, U. (2006). Induction of disease resistance by acibenzolar-S-methyl and o-hydroxyethylorutin against Botrytis cinerea in tomato plants Original Research Article. Crop Protection 25, 956-962.

McDowell, J. M., and Stacey, A. S. (2008). Molecular diversity at the plant–pathogen interface. Dev Comp Immunol 32, 736-744

Metraux, J. P., Signer, H., Ryals, J., Ward, E., Wyss-Benz, M., Gaudin, J., Raschdorf, K., Schmid, E., Blum, W., and Inverardi, B. (1990). Increase in salicylic acid at the onset of systemic acquired resistance in cucumber. Science 250, 1004-1006.

Millardet, P. R. A. (1885). Traitement du mildiou et du rot. J Agric Pratique 2, 513-516.

Millardet, P. R. A., and Gayon, U. (1888). Association française pour l'avancement des sciences : conférences de Paris. 17, Compte-rendu de la 17e session. Seconde partie. Notes et mémoires. Association française pour l'avancement des sciences Congrès (017 ; 1888 ; Oran, Algérie), 540-546. Available from: <http://catalogue.bnf.fr/ark:/12148/cb38809446f>.

Narusaka, Y., Narusaka, M., Horio, T., and Ishii, H. (1999). Comparison of local and systemic induction of acquired disease resistance in cucumber plants treated with benzothiadiazoles or salicylic acid. Plant Cell Physiol 40, 388-395.

Nurnberger, T., Brunner, F., Kemmerling, B., and Piater, L. (2004). Innate immunity in plants and animals: striking similarities and obvious differences. Immunol Rev 198, 249-266.

Pankin, A. A., Sokolova, E. A., Rogozina, E. V., Kuznetsova, M. A., Deahl, K., Jones, R. W., and Khavkin, E. E. (2010). Searching among wild Solanum species for homologues of RB/Rpi-blb1 gene conferring durable late blight resistance. Twelfth EuroBlight workshop Arras (France), 3-6 May 2010 Posters PPO-Special Report no 14 277 - 284.

Panter, S. N., and Jones, D. A. (2002). Age-related resistance to plant pathogens. Advances in botanical research *38*, 251-280.

Povero, G., Loreti, E., Pucciariello, C., Santaniello, A., Di Tommaso, D., Di Tommaso, G., Kapetis, D., Zolezzi, F., Piaggesi, A., and Perata, P. (2011). Transcript profiling of chitosan-treated Arabidopsis seedlings. J Plant Res DOI 10.1007/s10265-010-0399-1.

Qin, X. F., Holuigue, L., Horvath, D. M., and Chua, N. H. (1994). Immediate early transcription activation by salicylic acid via the cauliflower mosaic virus as-1 element. Plant Cell *6*, 63-874.

Raffaele, S., Farrer, R. A., Cano, L. M., Studholme, D. J., MacLean, D., Thines, M., Jiang, R. H., Zody, M. C., Kunjeti, S. G., Donofrio, N. M., *et al.* (2010). Genome evolution following host jumps in the Irish potato famine pathogen lineage. Science *330*, 1540-1543.

Ramey, B. E., Koutsoudis, M., von Bodman, S. B., and Fuqua, C. (2004). Biofilm formation in plant-microbe associations. Curr Opin Microbiol *7*, 602-609.

Rapilly, F. (2001). Champignons des plantes : les premiers agents pathogènes reconnus dans l'histoire des sciences. CR Acad Sci Paris, Sciences de la vie / Life Sciences *324*, 893-898.

Rasmussen, J. B., Hammerschmidt, R., and Zook, M. N. (1991). Systemic Induction of Salicylic Acid Accumulation in Cucumber after Inoculation with Pseudomonas syringae pv syringae. Plant Physiol *97*, 1342-1347.

Rossi, V., Giosuè, S., and T., C. (2009). Modelling the dynamics of infection caused by sexual and asexual spores during Plasmopara viticola epidemics. J Plant Pathol *91* 615-627.

Ryals, J., Weymann, K., Lawton, K., Friedrich, L., Ellis, D., Steiner, H. Y., Johnson, J., Delaney, T. P., Jesse, T., Vos, P., and Uknes, S. (1997). The Arabidopsis NIM1 protein shows homology to the mammalian transcription factor inhibitor I kappa B. Plant Cell *9*, 425-439.

Schirawski, J., Mannhaupt, G., Munch, K., Brefort, T., Schipper, K., Doehlemann, G., Di Stasio, M., Rossel, N., Mendoza-Mendoza, A., Pester, D., *et al.* (2010). Pathogenicity determinants in smut fungi revealed by genome comparison. Science *330*, 1546-1548.

Shennan, C. (2008). Biotic interactions, ecological knowledge and agriculture. Philos Trans R Soc Lond B Biol Sci *363*, 717-739.

Silverstein, R. M. (1981). Pheromones: background and potential for use in insect pest control. Science *213*, 1326-1332.

Simon, C., and Daniel, R. (2011). Metagenomic analyses: past and future trends. Appl Environ Microbiol *4*, 1153-1161.

Siqueira, J. F., Jr., and Rocas, I. N. (2009). Community as the unit of pathogenicity: an emerging concept as to the microbial pathogenesis of apical periodontitis. Oral Surg Oral Med Oral Pathol Oral Radiol Endod *107*, 870-878.

Spanu, P. D., Abbott, J. C., Amselem, J., Burgis, T. A., Soanes, D. M., Stuber, K., Ver Loren van Themaat, E., Brown, J. K., Butcher, S. A., Gurr, S. J., *et al.* (2010). Genome expansion and gene loss in powdery mildew fungi reveal tradeoffs in extreme parasitism. Science *330*, 1543-1546.

Tada, Y., Spoel, S. H., Pajerowska-Mukhtar, K., Mou, Z., Song, J., and Dong, X. (2008). Plant immunity requires conformational changes of NPR1 via S-nitrosylation and thioredoxins. . Science *321*, 952-956.

Theodorakopoulos, N., Govettoa, B., Industria, B., Massi, L., Gaysinski, M., Deleury, E., Mura, C., Marais, A., Arbiol, G., Burger, A., *et al.* (2011). Biology and Ecology of Biofilms formed by a plant pathogen Phytophthora parasitica: from biochemical Ecology to Ecological Engineering. Procedia Environ Sci. In press.

Tyler, B. M., Tripathy, S., Zhang, X., Dehal, P., Jiang, R. H., Aerts, A., Arredondo, F. D., Baxter, L., Bensasson, D., Beynon, J. L., *et al.* (2006). Phytophthora genome sequences uncover evolutionary origins and mechanisms of pathogenesis. Science *313*, 1261-1266.

Uknes, S., Dincher, S., Friedrich, L., Negrotto, D., Williams, S., Thompson-Taylor, H., Potter, S., Ward, E., and Ryals, J. (1993). Regulation of pathogenesis-related protein-1a gene expression in tobacco. Plant Cell *5*, 159-169.

Vallad, G. E., and Goodman, R. M. (2004). Systemic acquired resistance and induced systemic resistance in conventional agriculture. Crop Sci. *44*, 1920-1934.

Van Loon, L. C., and Van Strien, E. A. (1999). The families of pathogenesisrelated proteins, their activities, and comparative analysis of PR-1 type proteins. Physiological and Molecular Plant Pathology *55*, 85-97.

Vergne, E., Grand, X., Ballini, E., Chalvon, V., Saindrenan, P., Tharreau, D., Nottéghem, J. L., and Morel, J. B. (2010). Preformed expression of defense is a hallmark of partial resistance to rice blast fungal pathogen Magnaporthe oryzae. BMC Plant Biol *10:206*, doi:10.1186/1471-2229-10-206

Vernooij, B., Uknes, S., Ward, E., and Ryals, J. (1994). Salicylic acid as a signal molecule in plant-pathogen interactions. Curr Opin Cell Biol *6*, 275-279.

Vleeshouwers, V. G., Rietman, H., Krenek, P., Champouret, N., Young, C., Oh, S. K., Wang, M., Bouwmeester, K., Vosman, B., Visser, R. G., *et al.* (2008). Effector genomics accelerates discovery and functional profiling of potato disease resistance and phytophthora infestans avirulence genes. PLoS One *3*, e2875.

Walters, D., Walsh, D., Newton, A., and Lyon, G. (2005). Induced Resistance for Plant Disease Control: Maximizing the Efficacy of Resistance Elicitors. Phytopathology *95*, 1368-1373.

Ward, E. R., Uknes, S. J., Williams, S. C., Dincher, S. S., Wiederhold, D. L., Alexander, D. C., Ahl-Goy, P., Metraux, J. P., and Ryals, J. A. (1991). Coordinate Gene Activity in Response to Agents That Induce Systemic Acquired Resistance. Plant Cell *3*, 1085-1094.

Wei, Z. M., and Beer, S. V. (1996). Harpin from Erwinia amylovora induces plant resistance. VII International workshop on fire blight. Acta Hortic *411*, 223-225.

Wei, Z. M., Lacy, R. J., Zumoff, C. H., Bauer, D. W., He, S. Y., Collmer, A., and Beer, S. V. (1992). Harpin, elicitor of the hypersensitive response produced by the plant pathogen Erwinia amylovora. Science *257*, 85-88.

Weller, D. M., Raaijmakers, J. M., Gardener, B. B., and Thomashow, L. S. (2002). Microbial populations responsible for specific soil suppressiveness to plant pathogens. Annu Rev Phytopathol *40*, 309-348.

West, S. A., Diggle, S. P., Buckling, A., Gardner, A., and Griffin, A. S. (2007). The social lives of microbes. Annu Rev Ecol Evol Syst *38*, 53-77.

Whalen, M. C. (2005). Host defence in a developmental context. Mol Plant Pathol *6*, 347-360.

Whisson, S. C., Boevink, P. C., Moleleki, L., Avrova, A. O., Morales, J. G., Gilroy, E. M., Armstrong, M. R., Grouffaud, S., van West, P., Chapman, S., *et al.* (2007). A translocation signal for delivery of oomycete effector proteins into host plant cells. Nature *450*, 115-118.

Williams, P., Camara, M., Hardman, A., Swift, S., Milton, D., Hope, V. J., Winzer, K., Middleton, B., Pritchard, D. I., and Bycroft, B. W. (2000). Quorum sensing and the population-dependent control of virulence. Philos Trans R Soc Lond B Biol Sci *355*, 667-680.

Zhang, X. D., Francis, M. I., Dawson, W. O., Graham, J. H., Orbovic, V., Triplett, E. W., and Zhonglin, M. (2010). Over-expression of the Arabidopsis NPR1 gene in citrus increases resistance to citrus canker Eur J Plant Pathol *128*, 91-100.

Zipfel, C. (2009). Early molecular events in PAMP-triggered immunity. Curr Opin Plant Biol *12*, 414-420.

7

Comparative Study of the Mobility of Malathion, Attamix and Thiodan as Obsolete Pesticides in Colombian Soil

Rosalina González Forero
La Salle University
Colombia

1. Introduction

The organizations responsible for carrying out control and surveillance of pesticides in Colombia stored large quantities of them many years ago. They didn't know exactly the impact generated in that storage, but for the nature of the pesticides it is highly possible the incorporation of these hazardous substances in the soil, due the mobility throughout the leaching of the residue. This study worked with soils of the Colombian Agricultural Institute (ICA) because it is the organization in charge of the management of pesticides in Colombia. The research identified the mobility process by leaching of waste from the obsolete pesticides Thiodan, Attamix and Malathion, which according to previous researches (Duarte and Guerrero 2006 & Gonzalez 2011) these chemicals were stored in large quantities and the containers were deteriorated by weathering conditions.

In the investigation was simulated a spill at two concentrations: Event 1 (non diluted or commercial presentation), and Event 2 (agricultural recommended doses). Measures of the concentrations of leach pesticides at different depths into the soil (10, 17 and 25 cm) were determined using gas chromatography. The soils studied were taken from two sources Mosquera and Villavicencio branches of ICA.

2. Pesticides[1]

Malathion 57% EC: Malathion is an organophosphate insecticide that inhibits cholinesterase from insects. It is used worldwide to control a wide range of crop pests and control campaigns insect vectors that transmit diseases to humans: mosquitoes, bedbugs and mosquitoes. It is a brownish-yellow liquid that smells like garlic. (Proficol 2008), (ATSDR, 2008)

Molecular Formula: $C_{10}H_{19}O_6PS_2$

Molecular Weight: 330.36

Active ingredient: Malathion S1, 2-di (etoxycarbonil) ethyl O, O-dimetilfosforoditioato, 604 g per liter of formulation to 20 oC.

Attamix: Its active ingredient is Chlorpyrifos. Also the pesticide is known as Dursban 23, which belongs to the family of pesticides, insecticides and organophosphates. It is white

---

[1]This study was developed also by Ricardo Campos as a Co-Author (Faculty Member at La Salle University) and the research was funded by La Salle University

crystal like solid and strong aroma. In the United States, Attamix is used for residential use and it was allowed until 2000 when it was restricted by the United States Environmental Protection Agency (U.S. EPA). (ATSDR, 1997)

Molecular Formula: $C_9H_{11}Cl_3NO_3PS$

Chemical Name: O, O – diethyl O – (3,5,6 – trichloro – 2.Pyridinyl) phosphorothioate

Thiodan E.C.: Is a pesticide that smells like turpentine, but it does not burn. It is used to control insects on crops and non-food grocery, and as a wood preservative.

Chemical Name: 6,7,8,910,10 – hexachloro – 1, 5,5 a, 6,9,9 a – hexahydro – 6, 9 – methane – 2,4,3 – benzodioxatiepin – 3 – Oxia

3. Leaching and mobility tests

Leaching is defined as the removal of a substance in a solid phase by a liquid phase which is in contact with it. The determination of the toxic characteristics of waste depends on the analytical method and therefore cannot generalize results if they have not established criteria for evaluating mobility of toxic compounds by experimental analysis.

The factors or variables that limit the leaching method to make related to hazardous waste are given by:
- Surface area of waste
- Nature of the fluid extractor
- Value leachate / waste

From the variables mentioned above have been developed leaching tests classified according to the renewal of the exhaust fluid into two categories: the extraction tests and dynamic tests. The extraction tests were used in this research.

3.1 Mobility tests of a residue

The mobility test of a residue simulates infiltration conditions in the environment (soil). In the test: the waste infiltrates the soil and it can reacts with the components of that environment, causing public health risks due to environmental pollutants it absorbs; or do not react, but to infiltrate in large numbers so that the scope of such sources of groundwater. The methodology used for the mobility test was the procedure of the Colombian Institute Agustin Codazzi (IGAC) because it has a Reference Laboratories in this field in Colombia. It implies:

1. Sampling of the soil object of study.
2. Selection of specific method of sampling for the study. For the present study the sampling method can be seen in Figure 1. The method used was zig-zag due the number of samples to analyze at the laboratory.
3. Making a stripping of topsoil covering the ground.

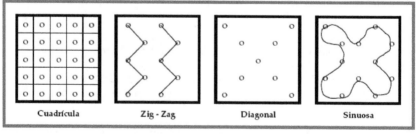

Fig. 1. Methodologies for soil sampling, Source. Codazzi. 2005.

3.2 Sampling for Mobility Analysis

To analyze the mobility, blocks of soil were taken from the fields of ICA. The soil block's dimensions were 25x25x25 cm. This height of the blocks was determined with a leaching test with water. In this test was spilled 1000 ml of water to calculate the deepest length.

The next set of pictures in figure 2 shows the process to take the samples of soils.

Fig. 2. Soil Sampling.

3.3 Experimental Units

The areas of those units, according to the heights determined in the field test for the assembly were: 25 x 25 cm, and a maximum depth of 25 cm with a free margin of 3 cm. All had the same dimensions and were evaluated at three different depths, being able to observe the behavior of the insecticide into the soil. The units were made in acrylic material to observe the movement of the pesticide. The lower surface was perforated for the collection of the leaching material. Figure 3 shows the experimental units.

Fig. 3. Experimental Units.

4. Simulated spill of the insecticide

As was mentioned in the previous section was performed spills for each of the events: Event 1 (non diluted or commercial presentation), and Event 2 (agricultural recommended doses). For each depth were spilled different volumes. They were obtained according to pilot tests to collect representative volumes. Below are the photographs of spills, as well as tables 1 to 6, which have the volumes applied and volumes collected after the spills for each pesticide.

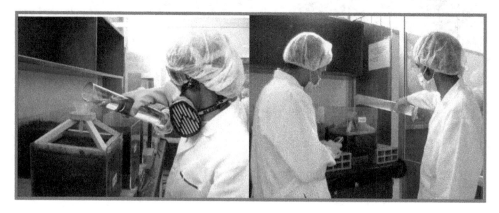

Fig. 4. Spill.

4.1 Malathion

Sample Height (cm)	Volume Applied (ml)	Volume Obtained (ml)
25	1000	32.8
25	1000	42.4
25	1000	54.5
17	680	45.6
17	680	73.4
17	680	58.7
10	400	51.4
10	400	46.8
10	400	68.1

Table 1. Event 1

Sample Height (cm)	Volume Applied (ml)	Volume Obtained (ml)
25	1000	48.6
25	1000	43.4
25	1000	63.7
17	680	72.4
17	680	64.2
17	680	56.6
10	400	76.5
10	400	70.4
10	400	62.8

Table 2. Event 2

4.2 Attamix

Sample Height (cm)	Volume Applied (ml)	Volume Obtained (ml)
10	1002	35,2
10	1002	9,5
10	1002	14,7
17	1003,4	21,5
17	1003,4	20,5
17	1003,4	16,8

Table 3. Event 1

Sample Height (cm)	Volume Applied (ml)	Volume Obtained (ml)
10	1002	43,3
10	1002	-
10	1002	20,5
17	1003,4	47,5
17	1003,4	-
17	1003,4	48,5

Table 4. Event 2

4.3 Thiodan

Sample Height (cm)	Volume Applied (ml)	Volume Obtained (ml)
25	400	142
25	400	152.5
25	400	115
17	680	95
17	680	45
17	680	92
10	1000	100
10	1000	18
10	1000	110

Table 5. Event 1

Sample Height (cm)	Volume Applied (ml)	Volume Obtained (ml)
25	400.8	110
25	400.8	46.5
25	400.8	98
17	681.36	200
17	681.36	177
17	681.36	164
10	1002	184
10	1002	240
10	1002	201

Table 6. Event 2

5. Initial concentrations

5.1 Malathion
The maximum concentration at which the insecticide is in the market is 650000 mg/L For the second event, was taken as the concentration the agricultural use specifications provided by the supplier, which was 604 mg/L.

5.2 Attamix
The commercial concentration of the pesticide had a value of 480000 mg/L. The concentration used for the second event was 960 mg/L of active ingredient chlorpyrifos.

5.3 Thiodan
The first event had a concentration of 350000 mg/L. The second event had as the concentration the value provided by the supplier as 0.7 mg/L.

6. Extraction

For all the samples obtained it was used a liquid - liquid extraction to isolate the pesticide from the samples and analyzed by gas chromatography. The methodology used was the described in the EPA 3510C method adjusted to the conditions of the Environmental Engineering Laboratory at the University of La Salle. The technique uses the dichloromethane (CH_2Cl_2) as a solvent and the process is characterized by separating the active ingredient of impurities and water content. To carry out the procedure first was stabilized the solution at pH 7.0 with sodium hydroxide; then, the solution was mixed with three portions of 90 ml of dichloromethane and the extract was separate in a gravity separation funnel. The figure 5 presents the extraction procedure:

Fig. 5. Extraction

6.1 Gas chromatography
The following chromatographic conditions were established for the analysis of the pesticides:
Injection volume: 2.0 mL (splitless mode)
Injector temperature: 280 ° C
Detector temperature: 280 ° C
Carrier gas, nitrogen at a constant flow of 1.0 mL / min.
Fuel Gas: Air: 300 mL / min.

Hydrogen: 30 mL / min
Make up gas: helium: 35 mL / min.
Oven temperature: 140 ° C start for 1 min. then a gradient of 20 ° C / min. 220 ° C and a stay at this temperature for 2 min. then again performed a gradient of temperature at 5 ° C / min. to 280 ° C with a stay at this temperature for 5 min. for a total analysis time of 24 min.

7. Results

The next tables present the results obtained from the chromatography analysis:

7.1 Malathion

Event	Height (cm)	$\overline{X}$ Concentration (mg/L)
1	25	581475
1	17	600825
1	10	615600
2	25	439.66
2	17	452
2	10	466

Table 7. Malathion Results

According to the above table, were constructed Figures 6 and 7; each of them corresponding to events 1 and 2. The correlation shows that the concentration decreases in proportion to the height of soil analyzed. The soil has a greater retention of material than the liquid phase, which implies that the impact will be much more evident in the solid phase than in the liquid. On the other hand the concentrations are quite high.

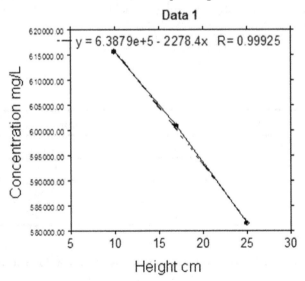

Fig. 6. Event 1

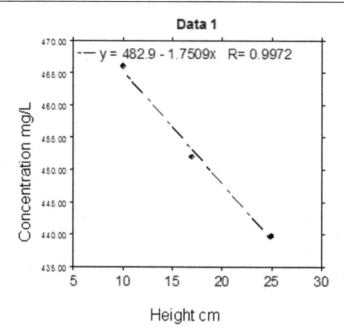

Fig. 7. Event 2

It is observed from the previous graphs that the degree of change in concentration of the pesticide in function to soil depth is significant. The graphs shows that the soil does not catch the pesticide enough. It means the pesticide has the ability to attack the biota present and generating highly toxic levels. For example, the LD50 is 1470 mg / kg oral rat, which implies that the current dose of stroke, 1 (one) liter of pesticide spilled, may die 600 rats of 600 g in weight. In the case of an adult rabbit weighing 4 kg, Malathion, which has a dermal LD50 of 5428, 28 animals die. Since the biotic point view these data have a strong importance.

7.2 Attamix

Event	Height (cm)	$\overline{X}$ Concentration (mg/L)
1	17	32.67
1	10	21.58
2	17	17.82
2	10	14.31

Table 8. Attamix Results

According to the above table were constructed the Figures 8 and 9, each corresponding to the different events. The concentration increases as the depth of soil sampled, quite contrary to the previous case. But the concentration is pretty low compared each event.

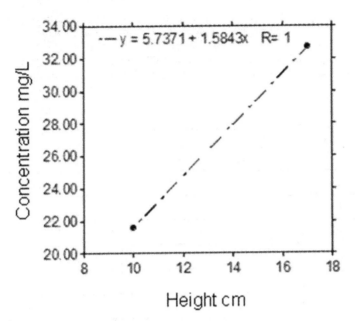

Fig. 8. Event 1

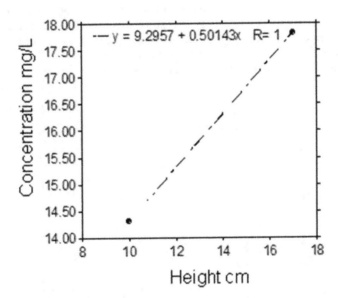

Fig. 9. Event 2

It is observed that the change in concentration of pesticide from events in function to soil depth is not significant. This suggests that the pesticide is highly related to soil because the pure pesticide leaching concentration is only almost twice as high concentration of diluted pesticide. According to its data sheet, this pesticide is highly toxic to aquatic organisms (LC50/EC50 <0.1 m/l in most sensitive species). The concentrations obtained by these two events generate mortality. For terrestrial organisms, where the pesticide is held, the first event with an approximately concentration of 900000 mg/L, the product is highly toxic to birds based on a diet (LC50 between 50 and 500 mg/L), causing their death.

7.3 Thiodan

Event	Height (cm)	$\overline{X}$ Concentration α-endosulfan (mg/L)	$\overline{X}$ Concentration β- endosulfan (mg/L)
1	25	63.3	39.10
1	17	45.9	24.93
1	10	30.9	13.6
2	25	5.29	4.3
2	17	11.11	6.84
2	10	19.88	12.02

Table 9. Thiodan Results

Thiodan has as active ingredients the- and β- endosulfan. According to the above table were constructed figures 10 and 11. The coefficients show that the concentration increases in the first event in proportion to the height of soil analyzed, meaning that deepens as the ground has been a greater release of material, which implies that the impact will be much more evident in the liquid phase in the solid, which implies an impact on water resources.

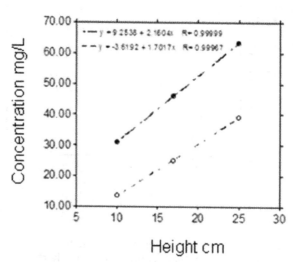

Fig. 10. Event 1. Black circle α- Transparent circle β-

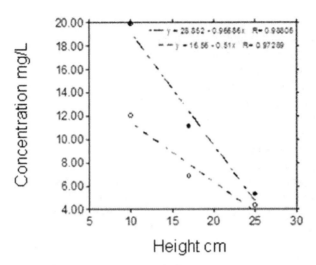

Fig. 11. Event 2. Black circle α- Transparent circle β-

The data from the second event presents a different phenomenon; it means that when the pesticide has more water content will favor retention is the solid phase as it delves into the less leach field.

Taking into account the concentration values reported in the leachate and the datasheet of the material, the lethal dose for biomarkers is 10 mg/kg. This indicates that all concentrations exceed the limits, except for the 25 cm diluted, so evidence that the potential risk of material, which in fact is prohibited in its entirety.

When is compared the leachate concentration with the initial concentration; is observed that the effect will be macro on soil because the material will be deposited into the soil. But the portion of leach is highly toxic yet.

8. Conclusions

Was evaluated the mobility of pesticide residues (Malathion, Thiodan And Attamix) in soils of the properties of ICA Mosquera and Villavicencio. For Malathion was found that the concentration decreases in proportion to the depth of the soil analyzed. It implies that the impact will be much more evident in the solid phase than in the liquid phase. The degree of change in concentration of the pesticide in function of the soil depth is representative, because the soil is not enough to catch all of the pesticide and it attacked the biota generating highly toxic levels.

In the case of Attamix the concentration increases as soil depth increases in a contrary way as the Malathion. But the retention of the pesticide is higher than from Malathion. It means that the leaching concentration is very lower than the initial concentration.

Were established concentrations of organochlorine and organophosphorus pesticides in the leachate showing differences between them. For the first event, Malathion about 600,000 mg/L, 28 mg/L for Attamix and 50 mg/L for Thiodan. For the second event 450 mg/L for Malathion, Attamix 15 mg/L and Thiodan and 10 mg/L. This indicates that Malathion is easily leached, while Attamix is preferably on the soil, as well as Thiodan.

9. References

Agency for Toxic Substances and Diseace Regitry. (Malathion CAS # 000121-73-5), 2008

ATSDR (Agency for Toxic Substances and Disease Resgistry). Clorpirifos. 1997. http://www.atsdr.cdc.gov/es/toxfaqs/es_tfacts84.pdf.

Chaves C. & Parra F. Estudio del impacto generado por un derrame simulado del insecticida Malathion en suelos de corpoica ubicados en el municipio de mosquera Cundinamarca, Bogota 2008. Pp. 20

Duarte, A. & Guerrero N. Evaluación de alternativas de manejo para eliminación de plaguicidas obsoletos existentes en el ICA; análisis de caso Villavicencio Mosquera". Bogotá 2006. p 44

Gonzalez R. Degradation of Organochlorine and Organophosphorus Pesticides by Photocatalysis: Chlorpiryfos and Endosulfan Case Study. 2011

Instituto Geográfico Agustín Codazzi. Suelos de Colombia, origen, evolución, clasificación, distribución y uso. Santafé de Bogotá. 1995

PROFICOL,2008, www.proficol.com.co/productos/pdf/insecticidas/Malathion%2057.pdf

8

Bioremediation of Hexachlorocyclohexane Contaminated Soil: Field Trials

H.K. Manonmani
Fermentation Technology and Bioengineering,
Central Food Technological Research Institute,
(Council of Scientific and Industrial Research)
India

1. Introduction

Hexachlorocyclohexane or the abbreviation HCH is identified as a monocyclic chlorinated hydrocarbon. HCH was discovered in 1825by Faraday, who just had discovered benzene. By reacting benzene with chlorine in bright sunlight, formation of HCH was observed. Neither Faraday nor the Dutch chemist Van der Linden, who in 1912 isolated the pure γ-isomer from a HCH mixture, realized the insecticidal potential of the compounds they produced (Amadori, 1993). The insecticidal properties of HCH were however first mentioned by Bender in a patent paper. HCH was first patented in the 1940s.. Dupire conducted detailed investigations on the insecticidal properties in 1940, and HCH was first used to combat the Colorado beetle (Stoffbericht, 1993). In 1942 Slade proved that γ-HCH was the sole carrier of the insecticidal properties of technical HCH (Stoffbericht, 1993). HCH production started commercially since 1947 in Germany. The common name of Hexachlorocyclohexane is "benzene hexachloride", which is incorrect according to the IUPAC rules (Galvan, 1999). Nevertheless it is still widely used, especially in the form of its abbreviation "BHC"

Technical HCH consists mainly of a mixture of various stereo-isomers, which are designated by Greek letters. Only one of these stereoisomers, γ-HCH, is the carrier of the insecticidal properties, while the other isomers are sometimes collectively referred to as "inactive isomers". The raw product from the chlorination of benzene contains about 14% γ-HCH and 86 % of inactive isomers, i.e. 65-70% α-, 7-10% β-,, 14-15% γ-, approximately 7% δ-,, 1-2% ε-HCH, and 1-2% other components. Therefore, in the production of one ton of technical HCH, 140 kg is γ-HCH and 860 kg is "inactive isomers". The latter is potentially waste and predestined for disposal. It is possible to extract and purify the active γ-HCH. If the purity is 99.0% or more it may be called "Lindane", which is an accepted common name for this substance. Lindane is also called γ-HCH, or γ- BHC and by FAO γ- BHC (technical grade). Technical grade HCH and fortified HCH (FHCH) containing a varying mixture of at least 5 isomers, with a minimum of 40% γ-isomer was available commercially. HCH is no longer produced in USA and few other European countries and cannot be sold for domestic use by EPA regulation as well as many other countries (FCH, 1984). However, in some developing countries HCH especially γ-isomer continues to be used because of economic purposes and also in public health programmes. Thus Technical grade HCH continues to be produced and all isomers except γ-isomer continue to be dumped unutilized. The production of Lindane creates huge amounts of isomers waste. The total quantity of waste will be about 8 times the Lindane output (Bodenstein, 1972), i.e. for each ton of Lindane produced 8 tons of waste will be generated. The large environmental consequences that are created can be imagined.

Photo from the mid-1990s of a temporary storage site for 200 000 tons of soil contaminated with waste HCH isomers.

Considering every ton of lindane produced generates approximately 6 -10 tons of other HCH isomers, a considerable amount of residues would be generated during the manufacture of this insecticide. For decades, the waste isomers were generally disposed off in open landfills like fields and other disposal sites near the HCH manufacturing facilities. After disposal, degradation, volatilization, and run off of the waste isomers occurred. If the estimate of global usage of lindane of 600,000 tons between 1950 and 2000 is accurate, the total amount of possible residuals (if it is assumed that a mean value of 8 tons of waste isomers are obtained per ton of lindane produced) amounts to possibly 4.8 million tons of HCH residuals that could be present worldwide giving an idea of the extent of the environmental contamination problem. Air releases of lindane can occur during the agricultural use or aerial application of this insecticide, as well as during manufacture or disposal. Also, lindane can be released to air through volatilization after application.

1.1 Fate of lindane
Once released into the environment, lindane can partition into all environmental media. Hydrolysis and photolysis are not considered important degradation pathways and reported half-lives in air, water and soil are: 2-3 days, 3-300 days and up to 2 - 3 years, respectively. A half-life of 96 days in air has also been estimated. Lindane can bioaccumulate easily in the food chain due to its high lipid solubility and can bio-concentrate rapidly in

microorganisms, invertebrates, fish, birds and mammals. The bio-concentration factors in aquatic organisms under laboratory conditions ranged from approximately 10 upto 4220 and under field conditions, the bio concentration factors ranged from 10 upto 2600.

Lindane is listed as a "substance scheduled for restrictions on use". This means that products in which at least 99% of the HCH isomer is in the γ-form (i.e. lindane, CAS: 58-89-9) are restricted to the following uses: 1. Seed treatment. 2. Soil applications directly followed by incorporation into the top soil surface layer 3. Professional remedial and industrial treatment of lumber, timber and logs. 4. Public health and veterinary topical insecticide. 5. Non-aerial application to tree seedlings, small scale lawn use, and indoor and outdoor use for nursery stock and ornamentals. 6. Indoor industrial and residential applications. Lindane is one of the listed priority hazardous substances for which quality standards and emission controls will be set at EU level to end all emissions within 20 years. Lindane is banned for use in 52 countries, restricted or severely restricted in 33 countries, not registered in 10 countries, and registered in 17 countries.

Lindane can be found in all environmental compartments, and levels in air, water, soil sediment, aquatic and terrestrial organisms and food. Humans are therefore being exposed to lindane as demonstrated by detectable levels in human blood, human adipose tissue and human breast milk in different studies in diverse countries. Exposure of children and pregnant women to lindane are of particular concern. γ-HCH has been found in human maternal adipose tissue, maternal blood, umbilical cord, blood and breast milk. Lindane has also been found to pass through the placental barrier. Direct exposure from the use of pharmaceutical products for scabies and lice treatment should be of concern. Exposure from food sources is possibly of concern for high animal lipid content diets and subsistence diets of particular ethnic groups. Occupational exposure at manufacturing facilities should be of concern, because lindane production implies worker exposure to other HCH isomers as well, for example the α- isomer is considered to be a probable human carcinogen.

Hepatotoxic, immunotoxic, reproductive and developmental effects have been reported for lindane in laboratory animals. The US EPA has classified lindane in the category of "Suggestive evidence of carcinogenicity, but not sufficient to assess human carcinogenic potential". The most commonly reported effects associated with oral exposure to γ-HCH are neurological. Most of the information is from case reports of acute γ-HCH poisoning. Seizures and convulsions have been observed in individuals who have accidentally or intentionally ingested lindane in insecticide pellets, liquid scabicide or contaminated food (WHO/Europe, 2003). Lindane is highly toxic to aquatic organisms and moderately toxic to birds and mammals following acute exposures. Chronic effects to birds and mammals measured by reproduction studies show adverse effects at low levels such as reductions in egg production, growth and survival parameters in birds, and decreased body weight gain in mammals, with some effects indicative of endocrine disruption.

Comparing to other POPs and hazardous waste problems, the HCH-residuals differ significantly as the extent of the problem is huge and as an environmentally sound disposal method will be necessary. However the enormous financial burden needed to achieve this will be a main barrier. On the other hand, the former practice of simple encapsulation is considered far from sustainable and will leave a huge number of time bombs in the global landscape. Hence bioremediation is the best option of removing these isomers from the contaminated environments. Microbial degradation of chlorinated pesticides such as HCH is usually carried out by using either pure or mixed culture systems. The main goal of the laboratory studies is to predict the biodegradation rates in the environment. But it is very difficult to extrapolate the results obtained in the laboratory systems to predict their fate in the environment [Spain et al 1990]. The microbial degradation of HCH isomers in liquid cultures has been studied using pure microbial cultures such as *Clostridium rectum*, *Pandoraea* [Ohisa et al 1990, Okeke et al 2002], mixed native soil microbial population (undefined consortium) [Bachmann et al 1988 a, Sahu et al 1995], *Phanerochaete chrysosporium* [Kennedy et al 1990, Mougin et al 1997] and sewage sludge under aerobic and anaerobic conditions [Bachmann et al 1988b, McTernan, and Pereira 1991, Buser, and. Mueller 1995]. The degradation of γ-and α-isomers was almost complete and β-and δ-isomers showed more resistance to degradation. At this stage it is imperative to develop technologies where all isomers of tech-HCH are degraded completely. In this communication we describe the degradation of tech-HCH in artificial plots and also in actual fields using a microbial consortium.

2. Materials and methods

2.1 Substrate
α-,β-,γ- and δ-HCH isomers (99% pure) were procured from Sigma–Aldrich Chemical Company, St. Louis, MO, USA. Technical grade HCH was obtained from Hindustan insecticides, Mumbai, India. Other chemicals and the reagents used in this study were of analytical grade and were purchased from standard chemical companies. *Rhizobium* and *Azospirillum* were obtained from IMTECH, Chandigarh. Soil used in small plot studies was collected from CFTRI campus without history of HCH usage.

2.2 Microbial consortium
The microbial consortium capable of degrading Tech-HCH was developed in our laboratory by long term enrichment of HCH contaminated soil and sewage according to Manonmani et al. [2000]. The Tech-HCH degrading consortium that got enriched was acclimated with increase in concentration of Tech-HCH from 5 to 25 ppm. The consortium thus obtained was maintained as liquid culture in minimal medium containing 10 ppm of Tech- HCH and in minimal agar medium containing 10 ppm of Tech- HCH.

2.3 Culture medium
Wheat bran hydrolysate used for growth of microbial consortium was prepared by acid hydrolysis of wheat bran. Minimal medium used in degradation studies consisted of KH_2PO_4,0.675g; Na_2HPO_4, 5.455g and NH_4NO_3, 0.25g and 1L of water.

2.4 Degradation of Tech-HCH using inoculum grown on different carbon sources
2.4.1 Biomass build-up and pre exposure of the inoculum
Individual isolates of the microbial consortium were grown in wheat bran hydrolysate/ peptone- glycerol medium and the cells were harvested after 72 h of growth, washed well and pre-exposed to 25ppm of individual isomers and Tech-HCH for 72 h separately with addition of individual isomers and Tech- HCH every 24 h. The induced cells were harvested by centrifugation at 10,000rpm, 4°C, 10min, washed well and used as inoculum.

2.4.2 Essentiality of individual members of the consortium for degradation of Tech-HCH

Degradation of higher concentration (25ppm) of Tech-HCH was taken up to test the performance of different combinations of the members as well as individual isolates of the HCH-degrading consortium. The combinations as given in Table 2 were tested for their ability to degrade tech-HCH. Samples were harvested after known period of incubation and analysed for residual HCH isomers.

Microbial combination used	Degradation (%)			
	α-isomer	B-isomer	γ-isomer	Δ-isomer
1	15.70	4.62	48.41	4.66
2	6.21	7.19	40.21	1.66
3	6.82	2.52	33.33	6.04
4	5.06	10.27	23.16	5.21
5	6.66	4.73	11.58	5.91
6	8.14	3.05	18.58	3.07
7	1.21	6.27	28.63	4.32
8	21.86	8.77	47.55	6.27
9	8.5	6.85	28.55	5.58
10	17.92	7.68	16.32	6.00
1-2	7.06	5.31	51.16	4.68
1-3	11.25	9.11	56.20	8.5
1-4	34.35	20.66	62.41	13.82
1-5	48.62	50.14	70.73	26.24
1-6	64.36	62.18	72.48	46.86
1-7	68.24	74.28	76.88	58.46
1-8	70.11	76.32	84.14	63.46
1-9	80.12	81.32	100	74.32
1-10	94.23	95.16	100	85.11

Table 2. Essentiality of individual members of the consortium for degradation of Tech-HCH

2.4.3 Preparation of inoculant formulations and their stability

Inoculant formulations containing the HCH-degrading microbial consortium and *Rhizobium* and *Azospirillum* were prepared using *Sphagnum* mass and wheat bran. The inoculum used contained 10^7 cells/g of the substrate. The consortium was mixed with *Sphagnum* mass. *Rhizobium* and *Azospirillum* were added to these separately. The inoculant formulation was prepared as both pellet and wet powder. The formulations were checked at regular intervals for survivability of the consortial members and their degradation capacity.

2.5 Degradation of Tech-HCH in different soils
Different soils such as clay soil, red soil, garden soil, soils from coconut, coffee, turmeric and tomato plantations were studied to evaluate the degradation of Tech- HCH. The substrate and inoculum were used at 25 μg g^{-1} and 500 μg protein g^{-1} soil respectively.
All the experiments were done in replicates of ten for each parameter studied.

2.6 Degradation of technical hexachlorocyclohexane in small artificial plots
Four artificial plots of 2ft x 2ft x 0.5ft were prepared. Red loamy non- sterile soil containing 1-1.5% organic carbon and 32% water holding capacity and pH 7.0, collected from CFTRI campus, was taken in these plots. Each plot was spiked with 25 ppm of tech- HCH and mixed well. Two plots were inoculated with 72 h HCH- induced microbial consortium. Two plots were maintained as abiotic (uninoculated) controls. All plots were kept wet by regular sprinkling of water and the soil in the plots was mixed regularly. Samples from each plot were collected at 10 different locations selected randomly at intervals of 24 h (around 5 g each). Collected soil was mixed thoroughly. One-gram soil was used for recording colony-forming units (cfu) and other set was used for residue analyses.

2.7 Degradation of HCH in artificial test plots (large size)
Individual isolates of the HCH- degrading consortium were grown individually in 25 L carbuoys, containing 20 L wheat bran hydrolysate medium (containing 0.75 % sugars) for 72 h. The cells were harvested and mixed at equal OD$_{600}$. The reconstituted consortium was induced with 25 ppm of Tech- HCH for 72 h in minimal medium, centrifuged, washed well in minimal medium and used as inoculum after resuspending the cells in known volume of minimal medium. Six artificial plots of 2m x 1m x 0.5ft were prepared and red loamy soil collected from CFTRI campus was taken in these plots. Each plot was spiked with 25 ppm Tech- HCH and mixed well. Four plots were inoculated with 72 h HCH- induced microbial consortium. The inoculum was added at levels containing 10^7 to 10^8 cells/ g soil. Two plots were maintained as uninoculated controls (containing only HCH). The soil was mixed well at regular intervals and was kept moist by regular sprinkling of water. At intervals of 24 h, samples were collected at 25 different areas of the plot (10 g each). The collected soil was mixed thoroughly and used for both the analysis of growth and residual substrate.

2.8 Degradation of Tech-HCH in actual fields
8 plots measuring 2m^2 were chosen at CFTRI campus (Plate 1). The plots were prepared by tilling and toeing. The individual members of the microbial consortium were grown separately in a 25 L carbuoy containing 20 L of wheat bran hydrolysate medium (containing 0.75% reducing sugars). After 72 h of growth the cells were harvested and pooled at equal OD$_{600}$. The reconstituted consortium was induced with 25 ppm of Tech- HCH for 72 h. Then the microbial mass was separated by centrifugation, washed well and resuspended in required quantity of minimal medium. Spiking of Tech- HCH to the soil was repeated for four times on alternative days. The final concentration of Tech- HCH added to the soil was 25 ppm. Six of the plots received Tech- HCH while two remaining were maintained as unspiked, uninoculated controls. Two plots were inoculated with microbial consortium and two more plots received the one-month-old inoculant formulation prepared using *Sphagnum* mass. The soil was mixed at regular intervals and kept wet by regular sprinkling of water. The inoculum at 0 h was added at the level containing 10^7- 10^8 cells. Sampling was done at 24 h intervals. 25 samples were removed randomly from each plot and analysed for residual HCH isomers and survivability of microbial consortium.

Plate 1. HCH degradation experiments in open plots/actual fields

2.9 Bioassay of the remediated soil

Seeds of *Raphanus sativus* (radish) and *Aeblumuschus esculantus* (ladies finger), members of *Brassicaceae* and *Malvaceae* which showed very high toxic effects of HCH towards seed germination were used as indicator plants to study the degradation of HCH in bioremediated soil. All the four bioremediated plots and the HCH- spiked but uninoculated controls and non- HCH- spiked controls were all sown with seeds of both radish and ladies finger. The seeds were allowed to germinate and grow completely. The plants were checked for growth, deformalities etc.

2.10 Degradation of other organochlorine pesticides by the microbial consortium

Soil was spiked with different organochlorine pesticides such as endosulphan, heptachlor, endrin, dieldrin at 10ppm level. Microbial consortium induced with respective organochlorine pesticide was used as inoculum. Pesticide spiked soil was inoculated with microbial consortium at inoculum 500 μg protein /g soil. Moisture was maintained at 15 % by sprinkling water daily. Randomized sampling (1g soil from 2 different locations) was done at every 3h interval and analysed for residual spiked pesticide.

2.11 Degradation of Tech-HCH in native soil

The Tech-HCH degrading consortium was inoculated to different types of native soils. The types of soil chosen were clay soil; soil from coconut fields; tomato fields; coffee plantations; turmeric fields; red soil and garden soil. These soils were spiked with Tech-HCH and mixed well. It was inoculated with microbial consortium at 500 μg protein /g soil. Moisture was maintained at 15 % by sprinkling water daily. Randomized sampling (1g soil from 2 different locations) was done at every 3h interval and analysed for residual spiked pesticide.

3. Analytical

3.1 Growth

Growth of the consortium was determined by estimating total protein in the biomass by modified method of Lowry et al according to Murthy et al (2007). Cells were harvested from a suitable quantity of culture broth, washed with minimal medium, suspended in 3.4 mL distilled water and 0.6 mL of 20% NaOH. This was mixed and digested in a constant boiling water bath for 10 min. Total protein, in cooled sample of this hydrolysate, was estimated by using Folin-Ciocalteau reagent. A total of 0.5 mL of the hydrolysate was taken in a clean test tube. To this was added 5.0 mL of Lowry's C. After 10 min 0.5 mL of Lowry's D [Folin-Ciocalteau reagent (1:2)] was added and mixed well. The colour was read at 660nm after 20.0 min of standing at room temperature, using a spectrophotometer (Shimadzu UV- 160A, Japan). Total amount of protein was computed using the standard curve prepared with BSA (Bovine serum Albumin).

Growth was also measured in terms of colony forming units (cfu) as described by Sahu et al (1995) using appropriately diluted broth.

3.2 Quantification of HCH

Residual HCH was quantified by Thin Layer Chromatography (TLC) and Gas Chromatography (GC).

3.2.1 Extraction of residual HCH

The soil samples (whole cups), were removed after required period of incubation, they were air dried and extracted thrice with equal volumes of dichloromethane. The three solvent extracts were pooled and passed through column containing sodium sulphate (anhydrous) and florisil. These fractions were concentrated at room temperature and resuspended in a known volume of acetone and used for analysis of residue.

3.2.2 Thin Layer Chromatography

Thin layer chromatography (TLC) was done using silica gel G TLC plates. Residual substrate samples dissolved in required quantity of acetone were spotted on TLC plates and these plates were developed in cyclohexane. The residual tech-HCH spots were identified after spraying the air-dried developed plates with O-tolidine in acetone. The residual substrate spots were delineated by marking with a needle and the area was measured. The concentration of residual substrate was computed from a standard plot of log concentrations *versus* square root of the area prepared for standard tech-HCH.

3.2.3 Gas chromatography

Concentrated solvent containing residual substrate was resuspended in HPLC grade acetone and gas chromatography was done using Chemito gas chromatograph (GC 1000). After appropriate dilution 1µL sample was injected into gas chromatograph equipped with ^{63}Ni detector and capillary column DB 5 (30m X 0.25 mm) packed with (5% phenyl)-methylpolysiloxane. The column, injector and detector were maintained at 230^0 C, 230^0 C and 320^0 C respectively with a flow rate of carrier gas IOLAR grade I nitrogen at 1mL/min. The recovery of HCH isomers ranged from 92 to 95% from mineral salts medium. All the data presented in this study are based on triplicate estimations.

Other organochlorine pesticides were also estimated by GC under same conditions.

4. Results and discussion

The study is meant to assess the efficiency of microbial consortium to degrade HCH isomers contained in soil. The optimized conditions obtained in our previous study (Murthy and Manonmani, 2007) were adopted in this present study to find out the applicability. The optimized parameters used were: inoculum level 500 µg protein g^{-1} soil, pH 7.5 and incubation temperature 30^0 C.

The soil used was red soil with no history of HCH applications. Soil had a particle size less than 0.5 mm. This size was chosen to provide high superficial area for interaction between HCH and microorganisms.

4.1 Microbial consortium

HCH-degrading microbial consortium developed by the long-term enrichment of the contaminated soil and sewage samples (Manonmani et al, 2000; Bidlan and Manonamani, 2002). This was reconstituted by mixing the different consortia having the ability to degrade α-,β-,γ- and δ-HCH isomers. This reconstituted consortium was acclimated with increasing concentration of Tech-HCH from 5ppm through 25 ppm. in shake flasks through three consecutive transfers every 24h. This consortium was used to study the degradation of Tech-HCH. The community structure of the consortium was identified by dilution plating technique. The bacterial of the consortial community were identified by biochemical tests and following Bergey's Manual of Determinative Bacteriology. The identification was confirmed by using Microbact Identification system. The consortium was found to be made of ten bacterial isolates consisting of seven *Pseudomonas* spp.; one species each of *Burkholderia*, *Flavobacterium* and *Vibrio* (Table 1)

Different carbon sources, both simple and complex, were used to optimize   biomass production. Molasses supported highest biomass production, followed by glucose, sucrose, rice straw extract supplemented with glucose, rice straw hydrolysate, nutrient broth and wheat bran hydrolysate. The biomass grown on different carbon sources, when inoculated to 25ppm of HCH, showed that biomass grown on wheat bran hydrolysate showed better HCH-degradation. Nearly 80 – 90 % of all the four isomers disappeared by 72h of incubation. Next was molasses grown inoculum, which showed 65 – 70 % of degradation (Fig 1).

Sl. No.	Bacterial isolate	No.
1	*Pseudomonas fluorescens* biovar II	T$_1$
2	*Pseudomonas diminuta*	T$_2$
3	*Pseudomonas fluorescens* biovar I	T$_3$
4	*Burkholderi pseudomallei*	T$_4$
5	*Pseudomonas putida*	T$_5$
6	*Flavobacterium* sp.	T$_6$
7	*Vibrio alginolyticus*	T$_7$
8	*Pseudomonas aeruginosa*	T$_8$
9	*Pseudomonas stutzeri*	T$_9$
10	*Pseudomonas fluorescens* biovar V	T$_{10}$

Table 1. Composition of the Tech-HCH-degrading microbial consortium

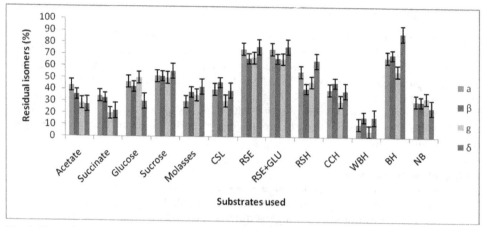

Fig. 1. Degradation of Tech-HCH with inoculum grown on different carbon sources
CSL , Corn steep liquor; RSE , Rice straw extract; RSH, Rice straw hydrolysate; CCH, Corn
cob hydrolysate; WBH, Wheat bran hydrolysate; BH, Bagassae hydrolysate; NB, Nutrient
broth.

4.2 Effect of induction/pre-exposure on the degradation of Tech-HCH

The pre-exposure of the HCH-degrading consortium was tried to understand the adaptation
of the developed consortium and to understand the choice of the inducer that could be used
for complete mineralization of tech-HCH. By exposing the WBH grown microbial
consortium to α-isomer resulted in the complete degradation of α-isomer of Tech-HCH by
72 h of incubation. β-, γ- and δ-isomers were still present at 2.035%, 6.019% and 24.13% levels
by the end of 72 h of incubation Similarly, β-, γ- and δ-HCH induced consortium degraded
the substrates used for induction better than other isomers. But, Tech-HCH induced
consortium degraded all the isomers (Fig 2) and in further experiments Tech-HCH induced
inoculum was used unless otherwise stated.

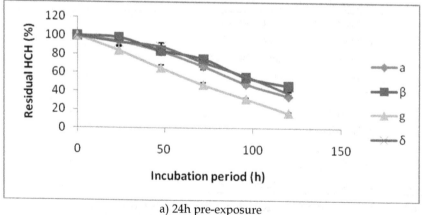

a) 24h pre-exposure

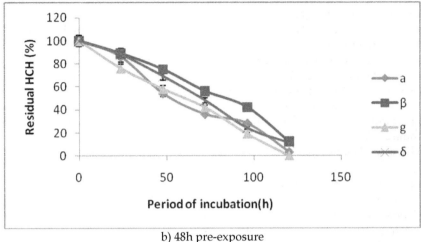

b) 48h pre-exposure

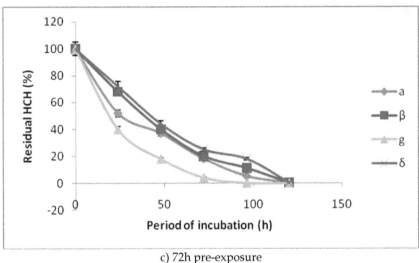

c) 72h pre-exposure

Fig. 2. Effect of Pre-exposure to Tech-HCH on Tech-HCH degradation

4.3 Essentiality of individual members of the consortium for degradation of Tech-HCH

Degradation of higher concentrations (25ppm) of Tech-HCH was taken up to test the performance of different combinations of the members as well as individual isolates of the HCH-degrading consortium. Table 2 describes the results of this study. It appears obvious that the presence of all the ten strains is necessary for the faster and efficient degradation of higher concentrations of HCH. With α-isomer as substrate, isolate T8 degraded 21% of the substrate by 120h. No other individual isolates were able to degrade this isomer. With increase in the number of individual members of the consortium the degradation of the α-isomer increased and when all ten isolates were mixed together 94% degradation of α-

isomer was observed. When β-isomer was used as a substrate, isolate T4 degraded up to 10% of the β-isomer. No other individual isolates were able to degrade this isomer up to 10 %. With increase in the number of individual members of the consortium the degradation of the β-isomer increased and when all ten isolates were mixed together 95% degradation of β-isomer was observed. When γ-isomer was used as a substrate, isolates T1, T8 and T2 were able to degrade 48, 47 and 49 % of γ-isomer respectively when inoculated individually. Combination of all T9 and T10 isolates could degrade the γ-isomer completely. With δ-isomer as sole substrate, isolate T8 showed highest degradation of 6 %. No other isolates were successful in degrading δ-isomer. Combination of all individual members of the consortium could degrade 85% of the isomer. Increase with addition of individual members showed increase in the degradation of δ-isomer.

4.4 Degradation of Tech-HCH in native soil

The Tech-HCH degrading consortium was inoculated to different types of native soils. Degradation was complete in all soils, but it was faster in red soil. In all soil types, γ-isomer disappeared faster (Fig 3). Degradation by native microflora was very low. The growth of the individual members of the consortium was also not inhibited by the presence of the native isolates.

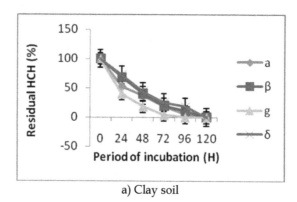

a) Clay soil

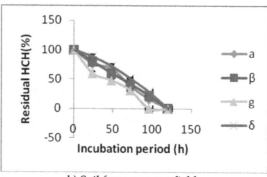

b) Soil from coconut fields

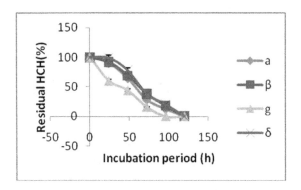

c) Soil from tomato fields

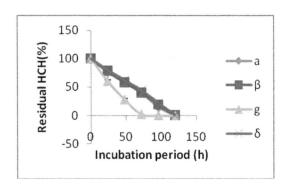

d) Soil from coffee plantations

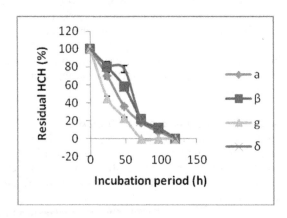

e) Red soil

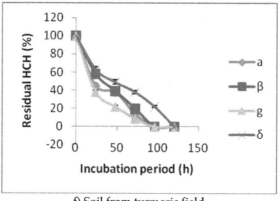

f) Soil from turmeric field

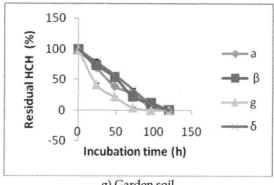

g) Garden soil

Fig. 3. Degradation of Tech-HCH in different native soils

4.5 Degradation of other organochlorine pesticides by the microbial consortium

The HCH degrading microbial consortium acclimated with both tech-HCH and different pesticides under study were inoculated separately to respective organochlorine pesticides. The analysis of soil at different periods of incubation time indicated that the degradation was good when the consortium induced with HCH was used. The organochlorine pesticides such as DDT, heptachlor, endrin, dieldrin and endosulphan disappeared with HCH induced consortium (Fig 4). With the consortium induced with respective substrates, all substrates, except endrin were degraded completely.

4.6 Preparation of inoculant formulations and their stability

Inoculant formulations containing the HCH-degrading microbial consortium and *Rhizobium* and *Azospirillum* were prepared using *Sphagnum* mass and wheat bran. The inoculum used contained 10^7 cells/g of the substrate. The inoculant formulation was prepared as both pellet and wet powder. The formulation was found to be stable for 60 days at refrigerated conditions. The formulations, when inoculated to soil containing 25 ppm tech-HCH, were found to degrade all the isomers of HCH completely (Fig 5). The formulation prepared in wheat bran took slightly longer time for complete degradation of tech- mixture.

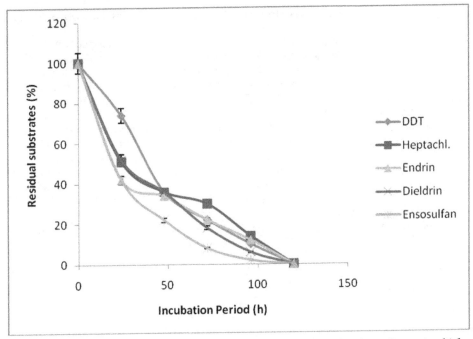

Fig. 4. Degradation of other organochlorine pesticides by Tech-HCH-degrading microbial consortium

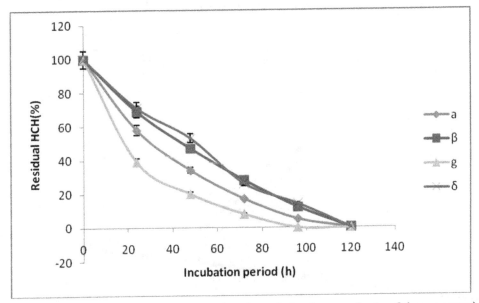

Fig. 5. Degradation of Tech-HCH by inoculant formulation (prepared using *Sphagnum* mass)

4.7 Degradation of technical hexachlorocyclohexane in small artificial plots of 2ft x 2ft x 0.5ft dimensions

Samples from each plot were collected at 10 different locations selected randomly at intervals of 24 h (around 5 g each). Collected soil was mixed thoroughly. One-gram soil was used for recording colony-forming units (cfu) and other set was used for residue analyses. The solvent fractions, passed through anhydrous sodium sulphate and florisil, were analysed for residual HCH isomers by GC. The maximum and minimum temperatures recorded during this period of study were 25- 28°C and 18- 22°C respectively. The degradation of Tech- HCH was complete by 120 h of incubation in inoculated plots (Fig.6). γ-Isomer was degraded faster followed by α-, β- and δ- isomers. The native microflora did not show the formation of any dead end metabolites or inhibition of degradation. The individual members of the consortium showed good survival during degradation. No competition or inhibition from native micro flora was observed towards both the substrate and the added microbial consortium (Table3). Uninoculated plots showed very little degradation of HCH- isomers. Only α- and γ- isomers were degraded by 2 and 4 % respectively, while β- and δ- isomers were not degraded by the native microorganisms even after 120 h of incubation.

!	Log of CPU									
	T1	T2	T3	T4	T5	T6	T7	T8	T9	T10
0	10.32±0.31	10.32±0.14	9.32±0.32	10.21±0.21	10.76±0.21	9.76±0.06	10.76±0.16	10.31±0.31	10.85±0.36	10.35±0.18
48	9.48±0.16	11.32±0.32	11.24±0.16	10.52±0.24	11.29±0.21	9.20±0.14	9.14±0.11	9.79±0.30	9.79±0.36	9.31±0.18
72	8.52±0.18	10.35±0.24	10.14±0.14	9.24±0.04	9.39±0.28	8.47±0.36	7.59±0.16	8.14±0.10	8.77±0.16	7.31±0.14
120	7.61±0.11	8.39±0.18	8.59±0.18	7.56±0.36	6.12±0.16	7.66±0.16	5.62±0.18	7.69±0.24	7.08±0.14	6.38±0.10

Table 3. Survivability of individual members of the consortium during the degradation of Tech- HCH

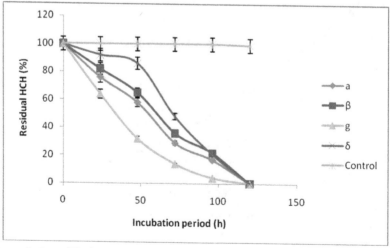

Fig. 6. Degradation of Tech-HCH in small artificial plots of 2ft x 2 ft x 0.5 ft

4.8 Degradation of HCH in artificial test plots of 2m x 1m x 0.5ft dimensions

At intervals of 24 h, samples were collected at 25 different areas of the plot (10 g each). The collected soil was mixed thoroughly and used for both the analysis of growth and residual substrate. It was observed that the degradation of all the isomers of HCH was complete with γ- isomer disappearing faster followed by α-, β-, and δ- isomers (Fig.7). The growth of the individual members of the consortium was also good (Table 4). The degradation of HCH-isomers was not observed by the native microflora.

!	Log of CPU									
	T1	T2	T3	T4	T5	T6	T7	T8	T9	T10
0	10.46±0.34	10.48±0.16	10.97±0.31	10.17±0.16	10.68±0.16	10.58±0.18	10.76±0.19	10.03±0.24	10.42±0.36	10.42±0.08
48	9.66±0.14	8.00±0.11	7.93±0.0.08	9.14±0.18	8.00±0.24	9.89±0.18	9.10±0.14	79.38±0.16	10.53±0.31	9.91±0.14
72	8.03±0.16	7.93±0.21	6.36±0.16	8.76±0.20	7.47±0.08	9.25±0.14	7.72±0.21	7.72±0.14	9.71±0.26	8.96±0.10
120	8.21±0.18	7.83±0.20	6.05±0.11	8.54±0.11	7.21±0.16	7.79±0.18	7.10±0.30	7.45±0.18	8.89±0.11	8.08±0.26

Table 4. Survivability of individual members of the consortium during the degradation of Tech-HCH

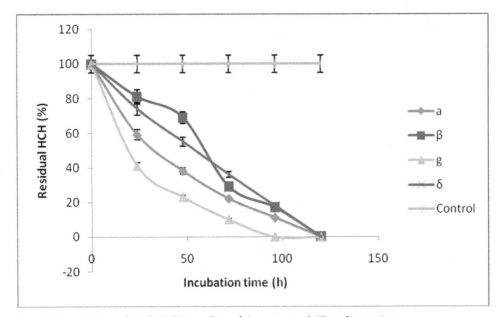

Fig. 7. Degradation of Tech-HCH in plots of 2m x 1 m x 0.15 m dimensions

4.9 Degradation of Tech-HCH in actual fields measuring 2m^2

25 samples were removed randomly from each plot. Collected soil from each plot was mixed well and 5 g soil from each sample was extracted. The solvent layers were pooled, passed through florisil and anhydrous sodium sulphate. The solvent fraction was pooled,

concentrated at room temperature and analysed for residual HCH. Compared to laboratory trials, the degradation of all isomers of HCH by the microbial consortium took 168 h for complete degradation (Fig.8). But the inoculant formulation took 10 days for complete degradation of all four isomers of HCH (Fig.9).

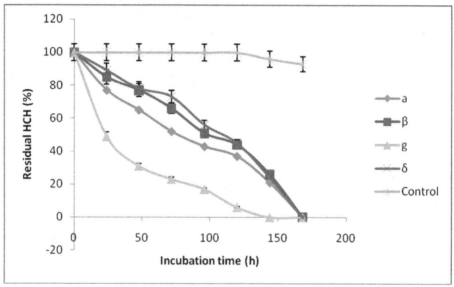

8a. Degradation by microbial consortium

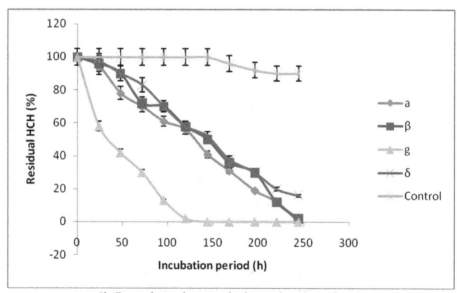

8b. Degradation by microbial inoculant formulation

Fig. 8. Degradation of Tech-HCH in actual open fields

4.10 Bioassay of the remediated soil

Seeds of *Raphanus sativus* (radish) and *Aeblumuschus esculantus* (ladies finger), members of *Brassicaceae* and *Malvaceae* which showed very high toxic effects of HCH towards seed germination were used as indicator plants to study the degradation of HCH in bioremediated soil. All the four bioremediated plots and the HCH- spiked but uninoculated controls and non-HCH- spiked controls were all sown with seeds of both radish and ladies finger. The germination of the plants was delayed in seeds sown in HCH- spiked soil. The height of the plants was reduced when compared to controls. Many other growth- related deformalities were also observed (Plate 2 and 3). In the bioremediated soil, growth related deformalities were not observed and the plant looked healthy *in par* with controls (Plate 4 and 5).

5. Discussion

HCH has been used worldwide as general broad spectrum insecticide for a variety of purposes including fumigation of the house hold and commercial storage areas, pest control on domestic animals, mosquito control and to eradicate soil-dwelling and plant-eating insects. Although only lindane has insecticidal property, HCHs as group are toxic and considered potential carcinogens (Walker et al., 1999) and listed as priority pollutants by the US EPA. Due to their persistence and recalcitrance, HCHs continue to pose a serious toxicological problem at industrial sites where post production of lindane along with unsound disposal practices has led to serious contamination. In addition, many countries including India have permitted HCH production (lindane is permitted to be used) and use. This has become a global issue due to problems of volatility and transportation of HCH isomers by air to remote locality (Galiulin et al., 2002; Walker et al., 1999). Due to the toxicity and persistence of HCH, soils contaminated with HCHs have been targeted for remediation. Biodegradation of α-, β-, γ- and δ- isomers of HCH have been extensively studied in the laboratory at individual level. But information is insufficient on pilot or full-scale *in situ* field settings. The HCH-isomers have been shown to differ in their persistence in soil and in their properties like solubility and volatility that determine their rates of biodegradation. Earlier studies suggested that degradation of HCH was faster under anoxic conditions and that microbial degradation was primary route of HCH disappearance from soil (MacRae et al., 1967). Microbial degradation of all the HCH-isomers has since been observed under oxic conditions both in soil (Bachmann et al., 1988 b); Doelman et al., 1985; Sahu et al., 1993) and in pure cultures of microorganisms (Bhuyan et al., 1993; Thomas et al., 1996). We have isolated in our laboratory a microbial consortium consisting of ten bacterial isolates which have got the capacity to degrade HCH (Manonmani et al., 2000; Murthy and Manonmani, 2007) under oxic conditions. Translation of the laboratory scale trials to small plots and actual open fields was studied in soil

The defined microbial consortium used in the degradation of tech-HCH was developed by the long-term enrichment of the contaminated soil and sewage samples. This microbial consortium was acclimated with increasing concentrations of tech-HCH from 5 to 25 ppm. The consortium that got established at 25 ppm level was used in our studies. The initial inoculum used in acclimation is obtained from diverse sources such as HCH contaminated soil and sewage. The advantage of sewage is that it provides sufficient inoculum during acclimation. As the test compound is used as a sole source of carbon and energy the organisms having the machinery for the degradation of the compound would survive and therefore would be able to accomplish the mineralization process. Moss [1980] employed an

a. Growth of radish in control soil.

b. Growth of radish in HCH- treated soil.

c. Growth of radish in Bioremediated soil.

Plate 2. Bioassay in soil

a. Growth of ladies finger in control soil.

b. Growth of ladies finger in HCH- treated soil.

c. Growth of ladies finger in Bioremediated soil.

Plate 3. Bioassay

a.   Growth of ladies finger in control soil.

b.   Growth of ladies finger in HCH- treated soil.

c.   Growth of ladies finger in Bioremediated soil.

Plate 4. Yields of Ladies finger in bioassay experiments

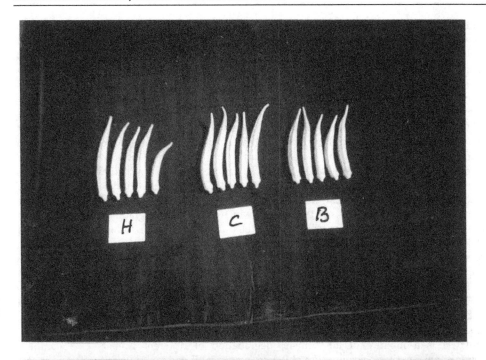

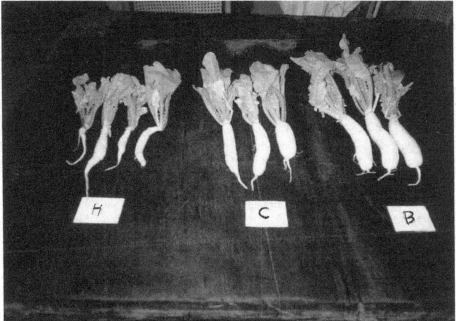

Plate 5. Yields of radish in bioassay experiments

acclimation and enrichment procedure that used a continuous culture of microorganisms growing at very low specific growth rates. The compound being tested was applied continuously at low concentrations and the concentration was increased in a systematic manner. Similar acclimation technique has been used by Bidlan and Manonmani [2002] to isolate DDT degrading microorganisms. Acclimation also would help in evading toxicity prior to the actual degradation and this is an essential part for further degradation studies. Acclimation would result in altered composition of the microbial populations involved in the early stages of degradation. Pre-exposure helped in obtaining faster degradation without any lag. The degradation appeared to have started as soon as inoculum and substrate were together. There was no initial lag in degradation with the inoculum induced with any of the isomers of HCH. In all these cases, the HCH-isomer used for induction was degraded completely and other isomers were partially degraded. This probably could be due to presence of required enzyme that got induced with the particular isomer. The partial degradation could be due to the multiple functions of the enzyme which was able to degrade the substrates only to certain extent. The partial degradation could be due to the enzymes, which might not have got induced by the other isomer used, or there could be inhibition of pathway enzymes by the intermediary metabolites formed during the degradation of non-inducer substrates. Also the complete degradation could be achieved with increase in incubation time. It has been reported that the degradation of α-and γ-isomers follow the same pathway and the degradation of β--HCH has been deciphered to only one or two steps of biodegradation (Nagata et al 2005). To our knowledge no reports are available on the degradation pathway of δ-HCH degradation. In our studies also the degradation of these isomers might be following a different pathway, As tech-HCH is a mixture of all isomers, degradation of different isomers is a complex phenomenon, as these enzymes might face many inhibitory/stimulating effects by the intermediates formed by different isomers present together. Failure to achieve results with consistent complete mineralization, on the other hand, would also suggest that complete biodegradation is not possible in short time or that it is dependent on co-metabolism. Thus, the tech-HCH induced inoculum was used in further studies. The microbial consortium developed in the laboratory was capable of degrading all isomers of HCH and the biodegradation could occur in a particular environment. With all the optimized conditions, the biodegradation of tech-HCH becomes a highly system specific event. These optimized results can be adapted well in the treatment of industrial effluent or water bodies contaminated with HCH. α- and γ-isomers have been reported to be degraded rapidly under aerobic and anaerobic conditions. α- and γ-isomers were degraded first in 12 h of incubation (Manonmani et al 2000 , Johri et al 1998 , Datta et al 2000), where as β-isomer was found to remain undegraded under similar environmental conditions (Bachmann et al 1988a, Beurakens et al 1991). The degradation of different isomers of HCH has been shown to be dependent on many features mainly the type of microorganism used, aeration, the adaptability of these microorganisms to the pollutants (Moreno and Buitron 2004), type of carbon source to cultivate them (Radha et al 2010), pre-exposure of the used organism to the pollutant (Bidlan and Manonmani 2002), etc. The recalcitrancy of the isomers also has been shown to play a key role in the degradation of different isomers of HCH (Bachmann et al 1988a, Haider and Jagnow 1975). In our studies also, the time required for the highly recalcitrant β- and δ-isomers was more compared to the other two isomers. The degradation of tech-HCH by the microbial consortium appears to be gratuitous metabolism where in, the substrate, i.e. tech-HCH is used as a sole source of carbon and energy and no other co-substrates are being supplemented. This is evident from the survival of all members of consortial community during degradation. However, no substantial growth was observed, i.e. the cells behaved as resting cells as the amount of carbon supplied by the substrate is not sufficient to support good growth of the consortium. But the cell count was being maintained during degradation. The initiation of degradation might be by the enzyme system that was already induced during pre exposure cycle, and hence degradation did not show any lag. The pH of the medium had a substantial effect on the survivability of the members of the consortial community. At low pH levels, there was decrease in the microbial population. The degradation was observed to take place over a wide range of pH from 4.0 to 8.0 (Murthy and Manonmani, 2007). Degradation was found to decrease at pH 9.0. The degradation of each isomer was influenced by the presence of other isomer. The HCH isomers, i.e. non-growth substrates have γ-isomer, the more easily degradable and β- and δ-isomer very highly recalcitrant. These two isomers thus show resistance toward degradation. It could be that because of recalcitrance, the structure may prevent it fitting into enzyme within the cell when it is likely to accumulate or the transformation product of one substrate may become toxic than the original substrate that might result in slower rate of degradation and also the degradative pathway of each isomer may be different which would be influenced by many factors. However, the different isomers present in the mixture will not associate in their co-metabolic degradation.

As our microbial consortium consisted of aerobic microorganisms, soil was mixed often in small plots to facilitate aeration. Similar degradation of α-HCH under oxic conditions in either moist soil or soil slurries has been reported (Doelman et al., 1985). Degradation of 23 mg kg^{-1} day^{-1} of α-HCH in soil under oxic conditions has been obtained (Bachmann et al., 1988a, b). However, reduction of 13 mg kg^{-1} day^{-1} was obtained under methanogenic conditions. Van Eekert et al. (1998) have reported the removal of α-HCH from a sandy soil containing low concentration of the isomer in slurries where lactate or sulfide had been added to reduce redox potential. Degradation of HCH was faster under anoxic conditions and that microbial degradation was primary route of HCH disappearance from soil (MacRae et al., 1967). Microbial degradation of all the HCH-isomers has since been observed under oxic conditions both in soil (Bachmann et al., 1988 (a,b); Doelman et al., 1985; Sahu et al., 1993) and in pure cultures of microorganisms (Bhuyan et al., 1993; Thomas et al., 1996). Soil slurry has been adopted for the microbial degradation of pesticides, explosives, polynuclear aromatic hydrocarbons, and chlorinated organic pollutants (Gonzalez et al 2003). In our studies translation of the laboratory scale trials to small reactors was studied in artificial plots and open actual fields. . The moisture content in soil has been shown to influence greatly HCH degradation. Chessells et al. (1988) have reported a correlation between soil moisture content and removal rates of HCH isomers in field agricultural soils. Enhanced removal of HCH in soils with higher moisture contents has been reported. This has been made possible due to prevailing anoxic conditions during flooding. Thus anaerobic metabolism has been reported to be existing in these soils. In our earlier studies, soil moisture content of 15 to 20% was found to give good biodegradation of HCH-isomers (unpublished data). This was used in the current study in small and actual plots. As our microbial consortium consisted of aerobic microorganisms, mixing at regular intervals was done to maintain oxic condition. Degradation of α-HCH in glass columns packed with contaminated sediments and held under methanogenic conditions has been reported (Middeldorp et al., 1996), although degrading population of microorganisms appeared not to be methanogens. Degradation of γ-HCH under oxic conditions has also been reported (Yule et al., 1967). β-HCH isomer, an indisputably most recalcitrant isomer, does not undergo biodegradaion easily. The concentration did not decrease noticeably in field study under any treatment (moist soil and oxic soil slurries in small pots) (Doelman et al., 1985)

Complete degradation of Tech-HCH was obtained in all the studies carried out. The bioassay of the bioremediated soil was carried out to check the degradation or mineralization of HCH isomers. The growth of Seeds of *Raphanus sativus* (radish) and *Aeblumuschus esculantus* (ladies finger) was poor in HCH spiked soil and their growth in bioremediated soil was *in par* with that in control soils. The crop yield in bioremediated soil was also *in par* with control soils.

6. Conclusion

The use of microbes to clean up polluted environments, bioremediation is rapidly changing and expanding the area of environmental biotechnology. Although much work is being done to remediate the polluted environment, our limited understanding of the biological contribution and their impact on the ecosystem has been an obstacle to make the technology more reliable and safer. In our studies a defined microbial consortium was able to degrade HCH (technical grade containing all four major isomers) up to 25 ppm level in soil at ambient temperature and neutral pH. The consortium was able to degrade HCH in artificial plots and also in open fields.

The inhibition of degradation by the presence of other isomers and native microflora was marginal. With the translation of lab trials to large scale trials coupled with process molecular microbiological techniques can make the bioremediation process more reliable and safer technology.

Although HCH removal has been observed under both oxic and anoxic bioremediation treatments, treatments under oxic condition have resulted in the almost complete removal of HCH *via* mineralization. These observations are on par with our results wherein under oxic conditions good degradation of HCH-isomers of technical mixture has been observed. Even though all the four isomers were present together in technical mixture, no adverse or inhibitory effects were observed by either parent compounds or their metabolites. We have tried to address the inadequately addressed topic of bioremediation of HCH contaminated soils in field studies. With the disadvantages of ex-situ bioreactors such as requirements for soil excavation, handling, conditioning and bioreactor construction/operation that typically increase treatment costs compared to most simple bioremediation techniques. The successful results obtained in small scale soil studies need to be addressed during translation further to still larger scale.

7. Legends to figures

Fig. 1. Degradation of Tech-HCH with inoculum grown on different carbon sources

CSL , Corn steep liquor; RSE , Rice straw extract; RSH, Rice straw hydrolysate; CCH, Corn cob hydrolysate; WBH, Wheat bran hydrolysate; BH, Bagassae hydrolysate; NB, Nutrient broth. Microbial consortium were grown in different carbon sources and induced with Tech-HCH and inoculated to Tech-HCH. Analysis was done as given under Methodology.
All the experiments were done in replicates of ten for each parameter studied.

Fig. 2. Preexposure and degradation of Tech-HCH

Individual isolates of the microbial consortium were grown in wheat bran hydrolysate/ peptone- glycerol medium and the cells were harvested after 72h of growth, washed well and preexposed to 25ppm of individual isomers and Tech-HCH for 72h separately with addition of individual isomers and Tech- HCH every 24h. The induced cells were harvested by centrifugation at 10,000rpm, 4°C, 10min, washed well and used as inoculum.

All the experiments were done in replicates of ten for each parameter studied.

Fig. 3. Degradation of Tech-HCH in different native soils:

Different soils such as clay soil, red soil, garden soil, soils from coconut, coffee, turmeric and tomato plantations were studied to evaluate the degradation of Tech- HCH. The substrate and inoculum were used at 25 µg g^{-1} and 500 µg protein g^{-1} soil respectively.

All the experiments were done in replicates of ten for each parameter studied.

Fig. 4. Degradation of other organochlorine pesticides by Tech-HCH-degrading microbial consortium.

Soil was spiked with different organochlorine pesticides such as endosulphan, heptachlor, endrin, dieldrin at 10ppm level. Microbial consortium induced with respective organochlorine pesticide was used as inoculum. Pesticide spiked soil was inoculated with microbial consortium at inoculum 500 µg protein /g soil. Moisture was maintained at 15 % by sprinkling water daily. Randomized sampling (1g soil from 2 different locations) was done at every 3h interval and analysed for residual spiked pesticide.

All the experiments were done in replicates of ten for each parameter studied.

Fig. 5. Degradation of Tech-HCH by inoculant formulation (prepared using Sphagnum mass)

Inoculant formulations containing the HCH-degrading microbial consortium and *Rhizobium* and *Azospirillum* were prepared using *Sphagnum* mass. two more plots received the one-month-old inoculant formulation prepared using *Sphagnum* mass. The soil was mixed at regular intervals and kept wet by regular sprinkling of water. Randomized sampling was done at every 3h interval and analysed for residual spiked pesticide.

Fig. 6. Degradation of Tech-HCH in small artificial plots of 2ft x 2 ft x 0.5 ft

Four artificial plots of 2ft x 2ft x 0.5ft were prepared. Red loamy non- sterile soil containing 1-1.5% organic carbon and 32% water holding capacity and pH 7.0, collected from CFTRI campus, was taken in these plots. Each plot was spiked with 25 ppm of tech- HCH and mixed well. Two plots were inoculated with 72 h HCH- induced microbial consortium. Two plots were maintained as abiotic (uninoculated) controls. All plots were kept wet by regular sprinkling of water and the soil in the plots was mixed regularly. Samples from each plot were collected at 10 different locations selected randomly at intervals of 24 h (around 5 g each). Collected soil was mixed thoroughly and was used for residue analyses.

Fig. 7. Degradation of Tech-HCH in plots of 2m x 1 m x 0.15 m dimentions.

Six artificial plots of 2m x 1m x 0.5ft were prepared and red loamy soil collected from CFTRI campus was taken in these plots. Each plot was spiked with 25 ppm Tech- HCH and mixed well. Four plots were inoculated with 72 h HCH- induced microbial consortium. The inoculum was added at levels containing 10^7 to 10^8 cells/ g soil. Two plots were maintained as uninoculated controls (containing only HCH). The soil was mixed well at regular intervals and was kept moist by regular sprinkling of water. At intervals of 24 h, samples were collected at 25 different areas of the plot (10 g each). The collected soil was mixed thoroughly and used for both the analysis of growth and residual substrate.

Fig. 8. Degradation of Tech-HCH in actual open fields:

8a. Degradation by microbial consortium.

8b. Degradation by microbial inoculant formulation:

8 plots measuring 2m² were chosen at CFTRI campus. The plots were prepared by tilling and toeing. Spiking of Tech- HCH to the soil was repeated for four times on alternative days. The final concentration of Tech- HCH added to the soil was 25 ppm. Six of the plots received Tech- HCH while two remaining were maintained as unspiked, uninoculated controls. Two plots were inoculated with microbial consortium and two more plots received the one-month-old inoculant formulation prepared using *Sphagnum* mass. The soil was mixed at regular intervals and kept wet by regular sprinkling of water. The inoculum at 0 h was added at the level containing 10⁷- 10⁸ cells. Sampling was done at 24 h intervals. 25 samples were removed randomly from each plot and analysed for residual HCH isomers.

Plate 1.

8 plots measuring 2m² were chosen at CFTRI campus.

Plate 2 and 3.

Seeds of *Raphanus sativus* (radish) and *Aeblumuschus esculantus* (ladies finger), members of *Brassicaceae* and *Malvaceae* which showed very high toxic effects of HCH towards seed germination were used as indicator plants to study the degradation of HCH in bioremediated soil. All the four bioremediated plots and the HCH- spiked but uninoculated controls and non- HCH- spiked controls were all sown with seeds of both radish and ladies finger.

Plates 4 and 5.

Seeds of *Raphanus sativus* (radish) and *Aeblumuschus esculantus* (ladies finger) grown in HCH-spiked, bioremediated and non-spiked soils.

8. References

Amadori, (1993) in The Legacy of Lindane HCH Isomer Production, International HCH & Pesticides Association, January 2006

A. Bachmann, P. Walet, P. Wijnen, W. de Bruin, J.L.M. Huntjens, W. Roelofsen, A.J.B. Zehnder, (1988a) Biodegradation of α- and β-hexachlorocyclohexane in a soil slurry under different redox conditions, Appl. Environ. Microbiol. 54 (1) 143–149.

A. Bachmann, D. de Bruin, J.C. Jumelet, H.H.N. Rijnaarts, A.J.B. Zehnder, (1988b) Aerobic bio mineralisation of α-hexachlorocyclohexane in contaminated soil, Appl. Environ. Microbiol. 54 (2) 548–554.

J.E. Beurakens, A.J. Stams, A.J. Zehnder,A. Bachmann, (1991)Relative biochemical reactivity of three hexachlorocyclohexane isomers, Ecotox. Environ. Saf. 21 (2) 128–136.

R. Bidlan and H.K. Manonamani, (2002), Aerobic degradation ofdichlorodiphenyltrichloroethane (DDT) by *Serratia marcescens* DT-1P, ProcBiochem. 38 (1) 49–56.

S Bhuyan,., B. Sreedharan, T. K. Adhya, and N. Sethunathan. (1993). Enhanced biodegradation of γ-hexachlorocyclohexane (γ-HCH) in HCH (commercial) acclimatized flooded soil: Factors affecting its development and persistence. Pestici. Sci., 38:49-55

Bodenstein, (1972) in The Legacy of Lindane HCH Isomer Production, International HCH & Pesticides Association, January 2006 de Bruin, (1979) in The Legacy of Lindane HCH Isomer Production, International HCH & Pesticides Association, January 2006

H. R. Buser, and M.D. Mueller, (1995), Isomer and enantioselective degradation of hexachlorocyclohexane isomers in sewage sludge under anaerobic conditions, Environ. Sci. Technol. 29 (3) 664–672.

Chessells, M. J., Hawker, D. W., Connell, D. W. and Papajcsik, I. A. (1988). Factors influencing the distribution of lindane and isomers in soil of an agricultural environment. Chemosphere, 17, 1741-1749

J. Datta, A.K. Maiti, D.P. Modak, P.K. Chakrabarty, P. Bhattacharya, P.K. Ray, (2000) Metabolism of gamma-hexachlorocyclohexane by *Arthrobacter citreus* strain B1-100: identification of metabolites, J. Gen. Appl. Microbiol. 46 (2) 59–67.

Doelman, P., L. Haanstra, E. de Ruiter, and J. Slange. 1985. Rate of microbial degradation of high concentrations of α-hexachlorocyclohexane in soil under aerobic and anaerobic conditions. Chemosphere, 14:565-570.

FCH, (1984), in   The Legacy of Lindane HCH Isomer Production, International HCH & Pesticides Association, January 2006.

R.V Galiulin, V. Bashkin, and R. A. Galiulina. (2002). Behavior of persistent organic pollutants in the air-plant-soil system. Water Air Soil Pollut., 137:179-191.

Galvan (1999). in   The Legacy of Lindane HCH Isomer Production, International HCH & Pesticides Association, January 2006

H. Haider, and G. Jagnow, (1975)Degradation of 14C and 36Cl labelled gamma-hexachlorocyclohexane by anaerobic soil microorganisms, Arch. Microbiol. 104 (2) 113-121.

A.K. Johri, M. Dua, D.Tuteja, R. Saxena, D.M. Saxena, R. Lal, (1998) Degradation of α, β, γ and δ-hexachlorocyclohexane by *Sphingomonas paucimobilis*, Biotechnol. Lett. 20 (90) 885–887.

D. Kennedy, S. Aust, and A. Bumpus, (1990) Comparative biodegradation of alkyl halide insecticides by the white rot fungus, *Phanerochaete chrysosporium* (BKM-F-1767), Appl. Environ. Microbiol. 56 (8) 2347– 2353.

I. C MacRae,.; K. Raghu, and T. F. Castro, (1967) Persistence and biodegradation of four common isomers of benzene hexachloride in submerged soils. J. Agric. Food Chem., 15, 911-914.

H.K. Manonmani, D.H. Chandrashekaraih, N.S. Reddy, C.D. Elcey, A.A.M. Kunhi, (2000)Isolation and acclimation of a microbial consortium for improved aerobic degradation of α-hexachlorocyclohexane, J. Agric. Food Chem. 48 (9) 4341–4351.

W.F. McTernan, and J.A. Pereira, (1991) Biotransformation of lindane and 2,4-D in batch enrichment cultures, Water Res. 25 (11) 1417–1423.

P. J.Middeldorp, M., M. Jaspers, A. J. B. Zehnder, and G. Schraa. (1996). Biotransformation of α-, β-, γ-, and δ-hexachlorocyclohexane under methanogenic conditions. Environ. Sci. Technol., 30:2345-2349.

L.P.Moss, (1980) The development of testing protocol to determine the biodegradability of pentachorophenol in activated sludge system, MSCE Thesis, Purdue University, West Lafayette,.

A.J. Moreno, and G. Buitron, (2004)Influence of the origin of the inoculum on the anaerobic biodegradability tests, Water Sci. Technol. 30 (7) 2345–2349.

C. Mougin, C. Peri Caud, J. Dubroca, and M. Asther, (1997) Enhanced mineralization of lindane in soils supplemented with the white rot basidiomycete *Phanerochaete chrysosporium*, Soil Biol. Biochem. 29 (9– 10) 1321– 1324.

H.M.R Murthy, and H.K .Manonmani, (2007). Aerobic degradation of hexachlorocyclohexane by a defined microbial consortium. J. Haz. Mat 149 (1): 18-25.

Y. Nagata, Z. Prokop, Y. Sato, P. Jerabek, Y. Ashwani Kumar, M. Ohtsubo, J. Tsuda, Damborsky, (2005) Degradation of ß-Hexachlorocyclohexane by Haloalkane Dehalogenase LinB from *Sphingomonas paucimobilis* UT26, Appl. Environ. Microbiol. 71 (4) 2183–2185.

N. Ohisa, M. Yamaguchi, N. Kurihara, (1980)Lindane degradation by cell free extracts of *Clostridium rectum*, Arch. Microbiol. 25 (3) (221.

B.C. Okeke, T. Siddique, M.C. Arbestain, W.J. Frankenberger, (2002) Biodegradation of gamma-hexachlorocyclohexane (lindane) and alpha hexachlorocyclohexane in water and soil slurry by *Pandoraea* species, J. Agric. Food Chem. 50 (9) 2548–2555.

S. Radha, M. Afsar, H.K. Manonmani, A.A.M. Kunhi, (2010) Effect of different isomers of hexachlorocyclohexane on the activity of microbial consortium and survival of its members, Asian J. Mirobiol. Biotechnol. Environ.Communicated).

M. Gonzalez,; Karina, S.B. and Julia, E. (2003) Occurrence and distribution oforganochlorine pesticides in tomato crops from organic production. J. Agric. Food Chem., *51*, 1353–1359.

S. K. Sahu, , K. K. Patnaik, S. Bhuyan, N. Sethunathan. (1993). Degradation of soil-applied isomers of hexachlorocyclohexane by a Pseudomonas sp. Soil Biol. Biochem., 25:387–391

S.K. Sahu, K.K. Patnaik, S. Bhuyan, B. Sreedharan, K. Kurihara, T.K. Adhya, N.Sethunathan, (1995) Mineralization of alpha-isomer, gamma-isomer andbeta-isomer of hexachlorocyclohexane by a soil bacterium under aerobicconditions, J. Agric. Food Chem. 43 (3) 833–837.

Stoffbericht, (1993) in The Legacy of Lindane HCH Isomer Production, International HCH & Pesticides Association, January 2006.

J.C. Spain, P.H. Pritchard, and A.W. Bourquin, (1980)Effects of adaptation onbiodegradation rates in sediment/water cores from estuarine andfreshwater environments, Appl. Environ. Microbiol. 40 726–734.

J. C. Thomas,., F. Berger, M. Jacquier, D. Bernillon, F. Baud-Grasset, N. Truffaut, P. Normand, T. M. Vogel, and P. Simonet. (1996). Isolation and characterization of a novel gamma-hexachlorocyclohexane- degrading bacterium. J. Bacteriol., 178: 6049-6055.

M. H. A Van Eekert,., N. J. P. Van Ras, G. H. Mentink, H. H. M.Rijnaarts, A. J. M. Stams, J. A. Field, and G. Schraa. (1998). Anaerobic transformation of β-hexachlorocyclohexane by methanogenic granular sludge and soil microflora. Environ. Sci. Technol., 32:3299-3304

K. Walker, D. A Vallero,. and R. G Lewis (1999). Factors influencing the distribution of Lindane and other hexachlorocyclohexanes in the environment. Environ. Sci. Technol,. 33 , 4373-4378.

WHO/Europe, (2003) World Health Organization (WHO). Lindane in Drinking Water: Background Document for Development of WHO Guidelines for Drinking-Water Quality.2004.
http://www.who.int/water_sanitation_health/dwq/chemicals/lindane/en/print.html

W. N Yule, M. Chiba, and H.V. Morley, (1967). Fate of insecticide residues; Decomposition of lindane in soil. J. Agric. Food Chem., 15 (6):1000-1004.

9

Application of Fenton Processes for Degradation of Aniline

Chavalit Ratanatamskul[1], Nalinrut Masomboon[2] and Ming-Chun Lu[3]
[1]Department of Environmental Engineering, Chulalongkorn University,
[2] Department of Interdisciplinary programs in Environmental Science, Graduate School,
Chulalongkorn University,
[3]Department of Environmental Resources Management, Chia-Nan University of
Pharmacy and Science,
[1,2]Thailand
[3]Taiwan

1. Introduction

Fenton process is one of advanced oxidation processes (AOPs) which are considered as alternative methods for treatment of non-biodegradable and toxic organic compounds. Fenton process has been widely used in the treatment of persistent organic compounds in water. In general, the mechanism of Fenton reaction included the formation of hydroxyl free radicals, which has E° of 2.8 V, that can oxidize and mineralize almost all the organic carbons to CO_2 and H_2O (Glaze et al., 1987), by the interaction of hydrogen peroxide with ferrous ions (Walling C., 1975). The Fenton's reagent is generally occurred in acidic medium between pH 2 -4 (Rodriguez et al., 2003). The advantage of Fenton process is the complete destruction of contaminants to harmless compounds, for instance, carbon dioxide and water (Neyen E. et al., 2003). However, its application has been limited due to the generation of the excess amount of ferric hydroxide sludge that requires additional separation process and disposal (Chang P.H., 2004). Therefore, electro-Fenton (EF) process is developed for minimizing the disadvantages of conventional Fenton process.

In the electro-Fenton method, the Fenton's regent was utilized to produce hydroxyl radical in the electrolytic cell, and ferrous ion was regenerated via the reduction of ferric ion on the cathode (Zhang et al., 2007). The regenerated ferrous ion will react with hydrogen peroxide and produce more hydroxyl radicals that can destroy the target compounds. However, the electro-Fenton reaction still faces several obstacles that must be overcome first such as the formation of ferric hydroxide sludge. Therefore, the new method which can promote the ferrous ion regeneration was focused in this part of experiment. The efficiency of pollutant removal and the reduction of ferric hydroxide sludge can be improved by using UV-radiation.

The photoelectro-Fenton process involves the additional irradiation of the solution with UVA light. Due to the generation of additional hydroxyl radical from the regeneration of ferrous ion and the reaction of hydrogen peroxide that reacted with UV light, so-called photoelectro-Fenton process (Brillas et al., 2000). Under UV-vis irradiation, the overall efficiency of the process increases due mainly to the regeneration of ferrous ions and formation of additional hydroxyl radicals. UVA light can favor (1) the regeneration of ferrous ion with production of more amount of hydroxyl radical from photoreduction of $Fe(OH)^{2+}$, which is the predominant ferric ion species in acid medium (Sun and Pignatello, 1993) and (2) the photodecomposition of complexes of ferric ion with generated carboxylic acids (Flox et al., 2007). The maximum adsorption wavelength of $Fe(OH)^{2+}$ species is less than 360 nm, visible irradiation may not drive the reaction of equation (1). An interesting and potentially useful modification of the photoreduction reaction takes advantage of the photo-lability of Fe(III)-oxalate complexes, which has efficiency up to 500 nm (Pignatello et al., 2006).

Aniline has been used as a target chemical. It is one of the main pollutants in wastewater, mainly from the chemical processes of dye, rubber industry, pesticide manufacture and pharmaceutical sectors (Song et al., 2007). In this study, the effects of the initial concentration of 2,6-dimethylaniline concentration, Fe^{2+} concentration, H_2O_2 dosage on 2,6-dimethylaniline and COD removal efficiency were explored. Moreover, this experiment intended to provide important information on the kinetic study of various Fenton processes (Fenton process, electro-Fenton process and photoelectro-Fenton process) on the aniline degradation.

2. Materials and methods

All chemicals used in this study were prepared using de-ionized water from a Millipore system with a resistivity of 18.2 MΩ cm^{-1}. Aniline, perchloric acid, ferrous sulfate and hydrogen peroxide were purchased from MERCK. Sodium hydroxide was purchased from Riedel-da Haën. All the preparations and experiments were realized at the room temperature. Synthetic wastewater containing 1 mM of aniline was dissolved with de-ionized water and then adjusted pH with perchloric acid. After pH adjustment, a calculated amount of catalytic ferrous sulfate was added as the source of Fe^{2+} in this experiment. Then, the H_2O_2 was added into the reactor. The electrical current (for electro-Fenton process and photoelectro-Fenton process) and UVA lamps (for photoelectro-Fenton process) were delivered through out the experimental period. The samples taken at predetermined time intervals were immediately injected into tubes containing sodium hydroxide solution to quench the reaction by increasing the pH to 11(Anotai et al., 2006). The samples were then filtered with 0.45 μm to remove the precipitates formed, and kept for 12 hours in the refrigerator before chemical oxygen demand (COD) analysis was conducted. This work has been carried out to investigate the effect of the concentration of hydrogen peroxide on the COD value.

The samples were also analyzed for aniline and COD. Aniline was analyzed by a gas chromatography (HP 4890II) equipped with a flame ionization detector (FID) and a SUPELCO Equity™ – 5 Capillary Column (length: 15m; id: 0.15 μm) with the rate 65 (°C/min), initial temperature 85°C and flow 10 psi. The solution pH was monitored using a SUNTEX pH/mV/TEMP (SP-701) meter. COD was measured by a closed-reflux titrimetic method based on Standard Methods (APHA, 1992). All experimental scenarios were duplicated. All experiments were carried out at batch mode using an acrylic reactor with a working volume of 5 liters. For the photoelectro-Fenton process, the anodes and cathodes used were mesh-type titanium metal coated with IrO_2/RuO_2 and stainless steel, respectively. The electrodes were connected to a Topward 33010D power supply operated at the desired electric current. The reactor was also equipped with two mixers to ensure appropriate agitation and the UVA source was turned on to initiate the reaction. The irradiation source was a set of twelve 0.6 W UVA lamps (Sunbeamtech.com) fixed inside a cylindrical Pyrex tube (allowing wavelengths $\lambda > 320$ nm to penetrate). In addition to all the experimental conditions mentioned above, UV light with maximum wavelength of 360 nm was irradiated inside the reactor, supplying a photoionization energy input to the solution of 7.2 W.

3. Results and discussion

In this step, the degradation of aniline was examined by various processes. Electrolysis, Photolysis, UV + hydrogen peroxide, Fenton, electro-Fenton, photo-Fenton and photoelectro-Fenton experiments in order to investigate the synergistic effect of Fenton's reagent combined with photo and electrochemical methods. As shown in Figure 1, the results show that electrolysis can remove 10% of 2,6-dimethylaniline within one hour. In the electrolysis method, aniline would be destroyed by reaction with adsorbed hydroxyl radical generated at the surface of a high oxygen-overvoltage anode from water oxidation. The same tendency can be found in the research of Brillas et al (Brillas et al., 1998).

Photolysis has lower degradation efficiency compared to electrolysis. The removal efficiency by photolysis was only 8% when using UVA lamps (12 lamps) at pH 3. The degradation of aniline was 12% when using UV+hydrogen peroxide after 60 minutes. Aniline was not well degraded by electrolysis, photolysis and UV+hydrogen peroxide. When using Fenton process, the degradation of aniline increased significantly compared to that when using direct photolysis, electrolysis and UV+hydrogen peroxide. For Fenton process, the degradation of aniline was 88% within one hour. This is due to the fast reaction of ferrous ion and hydrogen peroxide producing hydroxyl radicals (equation 1).

$$Fe^{2+} + H_2O_2 \rightarrow OH^{\bullet} + OH^- + Fe^{3+} \tag{1}$$

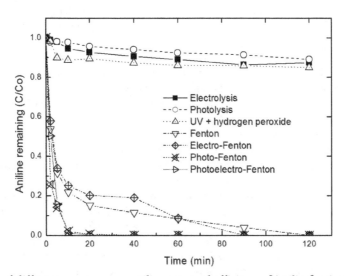

Fig. 1. Effect of different processes on aniline removal efficiency. [Aniline] = 1 mM; [Fe^{2+}] = 0.5 mM; [H$_2$O$_2$] = 20 mM; pH = 3; I = 1 A, UVA lamps = 12.

The 90% removal efficiency was achieved by the electro-Fenton process. The reason that electro-Fenton process can remove aniline more than the Fenton process is due to the regeneration of ferrous ion from equation (2) on the cathode side:

$$Fe^{3+} + e^- \rightarrow Fe^{2+} \tag{2}$$

and the ability of electricity that can produce hydroxyl radicals from water oxidation as described in equation (3).

$$H_2O \rightarrow OH^\bullet + H^+ + e^- \tag{3}$$

In photo-Fenton process, 100% of aniline degradation was observed at 60 min. Hence, photo-Fenton process is more efficient than Fenton process and electro-Fenton process. It is obvious that the photo-Fenton system enhanced the photooxidation of aniline. The hydroxyl radical which is a strong oxidant that can degrade aniline occurs by the following equation:

$$H_2O_2 + UV \rightarrow OH^\bullet + OH^\bullet \tag{4}$$

Degradation of aniline is mainly due to hydroxyl radical generated by photochemical reaction. In photo-Fenton process in addition to the above reaction the formation of hydroxyl radical (equation 1) also occurs by equation (4) and (5)

$$Fe^{3+} + H_2O + h\nu \rightarrow Fe^{2+} + OH^\bullet + H^+ \tag{5}$$

Meanwhile, the utmost removal efficiency was found when applied photoelectro-Fenton process, aniline was removed completely during the first 20 minutes. The degradation of aniline was due to the formation of hydroxyl radical from Fenton's reaction (equation 1) and the ferric ion would be reduced to ferrous ion from photoreduction (equation 5) and the cathode (equation 2). This would induce the formation of hydroxyl radicals efficiently.

The relative efficiencies of the above processes are in the following order: Photoelectro-Fenton > Photo-Fenton > Electro-Fenton > Fenton > UV + hydrogen peroxide > Electrolysis > Photolysis.

The degradation of aniline was monitored by measuring the COD removal. For COD removal, by electrolysis, photolysis, UV+hydrogen peroxide, Fenton, electro-Fenton, photo-Fenton and photoelectro-Fenton processes, it was followed the same trend as aniline degradation. The results revealed that electrolysis could remove COD only 9%, while photolysis was able to remove about 8% as shown in Figure 2. When UVA was combined with hydrogen peroxide, the COD removal increased to 14%. For Fenton process, the COD removal was 43% which is higher than using direct photolysis, electrolysis and UV+hydrogen peroxide .This is due to the fast reaction of ferrous ion and hydrogen peroxide producing hydroxyl radicals (equation 1).

The COD removal efficiency of aniline by the electro-Fenton process and photo-Fenton process were 49% and 52%, respectively. The reason might be due to the formation of hydroxyl radical from Fenton's reaction (equation 1) and the ferric ion would be reduced to ferrous ion from photoreduction (equation 5) in photo-Fenton process and the regeneration of ferrous ion on the cathode (equation 2) in electro-Fenton process. The highest COD removal was found when applied photoelectro-Fenton. It was about 70%. The decrease of COD can be attributed to the mineralization of aniline by the hydroxyl radicals from Fenton's reaction, from the electrochemically generated Fenton's reagent during electro-

Fenton and photoelectro-Fenton processes and the production of hydroxyl radical from photoreduction of Fe(OH)$^{2+}$ in photoelectro-Fenton process.

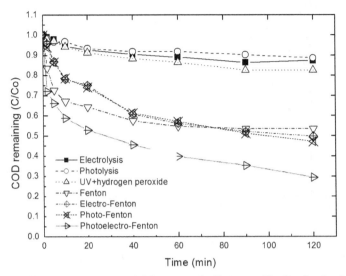

Fig. 2. Effect of different processes on COD removal efficiency. [Aniline] = 1 mM; [Fe^{2+}] = 0.5 mM; [H$_2$O$_2$] = 20 mM; pH = 3; I = 1 A, UVA lamps = 12.

From this part of experiment, the photoelectro-Fenton was found to be the most efficient treatment to oxidize aniline. Thus, the following part in this chapter will focused on this process.

3.1 Kinetic study of aniline degradation by photoelectro-Fenton process

The kinetic study of aniline degradation by the photoelectro-Fenton was investigated. In this study, the kinetic was performed using initial rate techniques (at the first 10 minute of the reaction) in order to eliminate the interferences from intermediates that might occur during the study period. The test range of each parameter was chosen according to the reality of photoelectro-Fenton process application and the needs of kinetic study in this experiment. The effects of the initial concentration of aniline, initial concentration of ferrous ion and hydrogen peroxide on the kinetic study of aniline degradation will be separately discussed in the following sections.

3.1.1 Effect of initial aniline concentration

The pollutant concentration is one of the important factors in photoelectro-Fenton process. The effect of initial aniline concentration on the removal efficiency of aniline by photoelectro-Fenton processes is shown in Figure 3. The figure clearly reveals that increasing the aniline concentration decreases the removal efficiency of aniline. When increasing the aniline concentration from 0.5 to 5 mM, the degradation of aniline decreased from 100%, 100% to 60%. The main reason for this phenomenon is the hydroxyl radical. Increasing in aniline concentration increases the number of aniline molecules, however, there is not enough hydroxyl radicals to degrade 2,6-aniline, then the removal efficiency decreases.

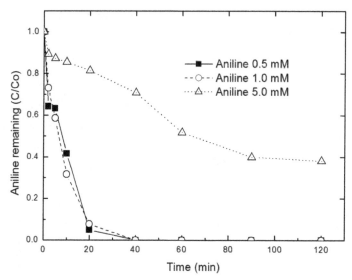

Fig. 3. Effect of initial aniline concentration on the removal efficiency of aniline by Photoelectro-Fenton process . [Fe²⁺] = 0.5 mM, [H₂O₂] = 20 mM, I = 1 A, pH = 3, UVA lamp = 12

The effect of the initial aniline concentration on the kinetic was studied by varying the initial concentration from 0.5 to 5 mM and at the experimental condition of 0.5 mM of ferrous ion, 20 mM of hydrogen peroxide, pH 3, an electric current of 1 amperes for and UVA 12 lamps. The study shows the aniline degradation kinetics at the given test condition. The second-order behavior appeared to fit the degradation of aniline by the photoelectro-Fenton process when plot the relationship between aniline degradation and initial aniline concentration was found in a good linear (R> 0.98).

| [Aniline] | Initial degradation rate |
| | Photoelectro-Fenton process |
(x10⁻³) M	(mMs⁻¹)
0.5	0.0006
1	0.0008
5	0.0010

Table 1. Initial rate of aniline degradation by varying initial concentration of aniline using photoelectro-Fenton process.

The results on Table 1 shows that when the aniline concentration was increased from 0.5 to 5 mM, the initial degradation rates increased. The high initial degradation rate at a high concentration of aniline was probably due to the excess amount of aniline to react with hydroxyl radical which produced during the process that destroyed the aniline.

For that reason, the initial rate for the higher concentration of aniline increased. These results are similar to the study of *p*-nitroaniline degradation by Fenton oxidation process by Sun and others (Sun et al., 2007) and the results are also similar to the removal of nitroaromatic explosives with Fenton's reagent (Liou and Lu, 2007).

For the effect of initial aniline on the removal efficiency, the results show that increasing the initial concentration of aniline will decrease the removal. The plot between the initial rate and aniline concentration showed a straight line with the slopes of 0.21 for the photoelectro-Fenton process. Therefore, the reaction rate equation became:

$$-\left(\frac{d[Aniline]}{dt}\right)_{photoelectro-Fenton} = r_{PEF,Aniline}\,[2,6-DMA]^{0.21} \tag{6}$$

where $r_{PEF,Aniline}$ is rate constant for the photoelectro-Fenton processes with respect to aniline.

3.2 Effect of ferrous ion concentration
The concentration of the catalyst is another important parameter for Fenton processes. The ferrous ion plays the catalyst's role in this Fenton's reaction. Normally, the rate of degradation increases with an increase in the concentration of ferrous ions. The effect of the ferrous ion concentration on the kinetic study of aniline degradation was studied by varying the ferrous ion concentration from 0.1 to 1 mM. The result showed that the removal efficiency of photoelectro-Fenton method was promoted when the ferrous ion concentration increased from 0.1 to 1 mM, as shown in Figure 4. The removal efficiency increased from 94% to 100 % in 40 minutes.

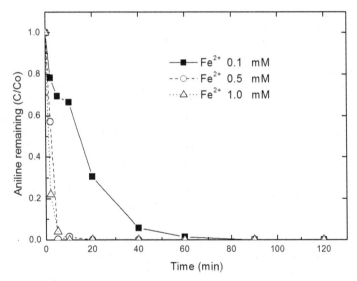

Fig. 4. Effect of initial ferrous ion concentration on the removal efficiency of aniline by Photoelectro-Fenton process . [Aniline] = 1 mM, [H_2O_2] = 20 mM, I = 1 A, pH = 3, UVA lamp = 12

For the initial degradation rate of aniline, the initial rate increased when the initial ferrous ion concentration increased from 0.0005 to 0.0016 mMs^{-1} in the photoelectron-Fenton process as listed in Table 2. Usually, when increasing the initial ferrous ion concentration, it will increase the generation rate of hydroxyl radical due to equation (1).

Therefore, this will enhance the degradation rate of aniline. Accordingly, the results underwent a similar direction. An improvement in removal efficiency by ferrous ion should also consider the supplement of hydrogen peroxide through out the experiment which might lead to more regeneration of ferrous ion from ferric ion. Therefore, the total regenerated hydroxyl radicals and the removal efficiency could increase as the ferrous ion concentration increased.

[Fe^{2+}] (x10^{-3}) M	Initial degradation rate Photoelectro-Fenton process (mMs^{-1})
0.1	0.0005
0.5	0.0014
1.0	0.0016

Table 2. Initial rate of aniline degradation by varying initial concentration of ferrous ion using photoelectro-Fenton process.

In photoelectro-Fenton process, the applied electricity and UVA lamps during experiment will enhance the regeneration of ferrous ion from ferric ion via equation (3) and (5). The supplied electrons from electrical current can regenerate ferrous ions rapidly and, thus, can react with hydrogen peroxide as long as hydrogen peroxide is still available in the reactor (Anotai et al., 2006). For that reason, the degradation of aniline continuously proceeded without the inadequacy of ferrous ions in the solution which in turn increased the efficiency of hydroxyl radical production.

The aniline degradation kinetics at the given test conditions was second–order behavior. The degradation kinetic for the photoelectro-Fenton process when plot the relationship between aniline degradation and aniline concentration was in a good linear (R> 0.95) when using second-order kinetic model. It was seen from the results that the rate of aniline degradation increased with the increase of ferrous ions concentration. The relationship between the initial rate and the ferrous ion concentration in the photoelectron-Fenton process was determined. From the calculation, it showed that the slope of photoelectro-Fenton processes was 0.5289. Thus, the kinetics for aniline degradation on the effect of ferrous ion concentration can be described by following equation:

$$-\left(\frac{d[Aniline]}{dt}\right)_{photoelectro-Fenton} = r_{PEF,Fe^{2+}} \ [Fe^{2+}]^{0.5289} \qquad (7)$$

where $r_{PEF,Fe^{2+}}$ is the rate constant for photoelectro-Fenton process with respect to the ferrous ion. The results indicated that ferrous ions play an important role in degradation of aniline by reacting with hydrogen peroxide to generate hydroxyl radicals. However, higher ferrous ion concentration supplementation by using photoelectro-Fenton processes is not recommended due to the large amount of ferric hydroxide sludge that could occur.

3.3 Effect of hydrogen peroxide concentration
Initial concentration of hydrogen peroxide plays an important role in the photoelectro-Fenton process. It is an oxidizing agent in the Fenton reaction. It has been observed that the percentage degradation of the pollutant increases with an increase in the concentration of

hydrogen peroxide (Pignatello, 1992). The effect of initial hydrogen peroxide concentration on the removal efficiency and kinetic rate of aniline degradation was investigated by varying the initial hydrogen peroxide concentration from 1 to 40 mM under the experiment condition of initial aniline concentration 1 mM, ferrous ion 0.5 mM, pH 3, electrical current 1 A and UVA 12 lamps.

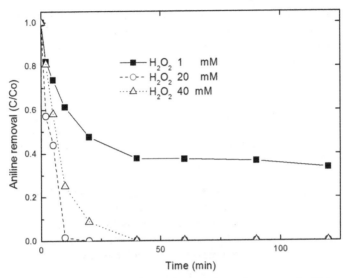

Fig. 5. Effect of initial hydrogen peroxide concentration on the removal efficiency of aniline by Photoelectro-Fenton process . [Aniline] = 1 mM, [Fe²⁺] = 0.5 mM, I = 1 A, pH = 3, UVA lamp = 12

It can be seen from the results that the removal efficiency of photoelectro-Fenton processes increased when increasing the initial concentration of hydrogen peroxide as shown in Figure 5. The removal efficiency increased from 63 % to 100 % as the hydrogen peroxide concentration increased from 1 to 40 mM in 60 minutes. This was from the production of hydroxyl radicals produced during the experiment by the reaction of ferrous ion and hydrogen peroxide in the solution. Similar result was obtained by Lin et al.,1999.

$[H_2O_2]$	Initial degradation rate
	Photoelectro-Fenton process
$(x10^{-3})$ M	(mMs⁻¹)
1	0.0006
20	0.0015
40	0.0012

Table 3. Initial rate of aniline degradation by varying initial concentration of hydrogen peroxide using photoelectro-Fenton process.

Table 3 shows that the initial rate of aniline degradation increased when the hydrogen peroxide concentration was increased from 1 to 20 mM. The initial rate increased from 0.0006 to 0.0015 mMs⁻¹ by photoelectro-Fenton process. This increase in the initial rate was due to the availability of hydrogen peroxide to react with ferrous ions in the solution. However, with the continuous increase in the initial hydrogen peroxide concentration to 40 mM, the increase of hydrogen peroxide leads to the decline of the initial rate. This phenomenon was probably due to the scavenging of hydroxyl radicals by hydroxyl radical as described via equation (5) (Lu et al., 1999). The accumulation of hydroperoxyl radicals also consumed hydroxyl radicals (Sun et al., 2007; Kang et al., 2002).

It was found that the second-order kinetic model is applicable to the aniline degradation quite well under photoelectro-Fenton process with R> 0.98. The relationship between the initial rate and hydrogen peroxide is linear and the slope is 0.2232 for the photoelectro-Fenton process, therefore:

$$-\left(\frac{d[Aniline]}{dt}\right)_{photoelectro-Fenton} = r_{PEF,H_2O_2} [H_2O_2]^{0.2232} \tag{8}$$

where r_{PEF,H_2O_2} is rate constant for the photoelectro-Fenton methods with respect to hydrogen peroxide.

3.4 The overall reaction rate equation for aniline degradation by photoelectro-Fenton

From the previous sections, the reaction rate equation of aniline degradation is proposed and it varied with the respect to aniline, ferrous ion and hydrogen peroxide concentrations in photoelectro-Fenton process. The overall degradation kinetic for aniline by the photoelectro-Fenton process can be summarized as shown below:

$$-\left(\frac{d[Aniline]}{dt}\right)_{photoelectro-Fenton} = r_{PEF}[Aniline]^{0.21} [Fe^{2+}]^{0.5289} [H_2O_2]^{0.2232} \tag{9}$$

where r_{PEF} is the overall rate constant for the photoelectro-Fenton method. From the equation (9), it can be seen that the degradation rate of aniline by photoelectro-Fenton process depended on Fenton's reagent both ferrous ion and hydrogen peroxide. However, from equation (9), the reaction shows that the aniline degradation, when applied with an electrical current and UVA lamp, the degradation rate is still depend on hydrogen peroxide. Moreover, the ferrous ion is also the important species for the degradation of this chemical so the Fenton's reagent still be the major key for this process.

The r_{PEF} , from equation (9) can be calculated using a non-linear least squares method which minimizes the sum of the error squares between the observed initial rates attained from the study and from the calculated initial rates. Accordingly, the r_{PEF} can be proposed by calculated using concentration of aniline, ferrous ion and hydrogen peroxide in millimolar (mM) units. From the calculation, the r_{PEF} was 1.32. Therefore, the final reaction rate equations can be described as:

$$-\left(\frac{d[Aniline]}{dt}\right)_{photoelectro-Fenton} = 1.32[Aniline]^{0.21} [Fe^{2+}]^{0.5289} [H_2O_2]^{0.2232} \tag{10}$$

4. Conclusions

Fenton, electro-Fenton and photoelectro-Fenton processes were able to oxidize aniline. However, in presence of electrical current and UVA lamps, the removal of aniline was higher. The results show that relative efficiencies of the AOPs processes on aniline degradation are in the following order: Photoelectro-Fenton > Photo-Fenton > Electro-Fenton > Fenton > UV + hydrogen peroxide > Electrolysis > Photolysis. These processes could degrade this compound. The photoelectro-Fenton process was found to give the highest degradation efficiency than the other processes. 100% degradation of 1 mM aniline was achieved in 40 minutes when 1 mM of ferrous ions, 20 mM of hydrogen peroxide, pH 3, 1 A of electric current and 12 lamps of UVA were applied. This was due to the regeneration of ferrous ion from ferric ion that can improve the performance of overall degradation. The overall rate equations for the degradation of aniline by photoelectro-Fenton process was evaluated as shown in equation (10). The kinetics established in this experiment was mathematically determined by considering the three important parameters including aniline, ferrous ion and hydrogen peroxide. The initial rate of aniline was increased with the increase of aniline concentration in all processes and increased with the increase of initial ferrous ion and hydrogen peroxide concentration. It can be indicated that aniline, ferrous ion and hydrogen peroxide have strong influences on the kinetic rate constants for aniline degradation.

5. Acknowledgements

This work was financially supported by The 90TH Anniversary of Chulalongkorn University Fund (Ratchadaphiseksomphot Endowment Fund) and The National Science Council of Taiwan (Grant: NSC95-2211-E-041-019).

6. References

[1] Anotai J, Lu MC, Chewpreecha P (2006) Kinetics of aniline degradation by Fenton and Electro-Fenton Processes. Water Research 40: 1841-1847
[2] APHA (1992) Standard Methods for the Examination of Water and Wastewater, 18th Ed.
[3] Brillas E, Mur E, Sauleda R, Sanchez L, Peral J, Domenech X, Casado J (1998) Aniline mineralization by AOP's: anodic oxidation, photocatalysis,electro-Fenton and photoelectro-Fenton processes. Applied Catalysis B: Environmental. 16(1):31–42.
[4] Brillas E, Calpe J.C, Casado J (2000) Mineralization of 2,4-D by advanced electrochemical oxidation processes. Water Research, 34, 2253-2262.
[5] Chang PH (2004) Treatment of Non-Biodegradable Wastewater by Electro-Fenton Method. Water Science and Technology 49(4): 213-218
[6] Flox C., Garrido J.A., Rodríguez R.M., Cabot P.I.., Centellas F., Arias C., Brillas E. (2007) Mineralization of herbicide mecoprop by photoelectro-Fenton with UVA and solar light. Catalysis Today, 129, 29-36.
[7] Glaze W.H., Kang J.W., Chapin R.H. (1987) Chemistry of water treatment processes involving ozone, hydrogen peroxide and ultraviolet radiation. Ozone Science and Engineering. 9, 335-352.
[8] Kang N, Lee D.S, Yoon J. (2002) Kinetic Modeling of Fenton Oxidation of Phenol and Monochlorophenols. Chemosphere. 47 : 915-924.

[9] Lin SH, Lin CM, Leu HG (1999) Operating characteristics and kinetic studies of surfactant wastewater treatment by Fenton oxidation. Water Research 33: 1735.

[10] Liou M-J and Lu M.C. (2007)Catalytic degradation of nitroaromatic explosives with Fenton's reagent. Journal of Molecular Catalysis A: Chemical. 277:155-163.

[11] Lu MC, Chen JN, Chang CP (1999) Oxidation of dichlorvos with hydrogen peroxide using ferrous ions as catalyst. Journal of Hazardous Materials B65: 277-288.

[12] Neyen E, Baeyens J (2003) A review of classic Fenton's peroxidation as an advanced oxidation technique. Journal of Hazardous Materials B98: 33-50.

[13] Pignatello J.J., Oliveros E., Mackay A. (2006) Advanced oxidation processes for organic contaminant destruction based on the Fenton reaction and related chemistry. Critical Review in Environmental Science and Technology 36, 1-84.

[14] Rodriguez ML, Timokhin VI, Contreras S, Chamarro E, Esplugas S (2003) Rate equation for the degradation of nitrobenzene by 'Fenton-like' reagent". Advances in Environmental research 7: 583-595.

[15] Song S., He Z., Chen J. (2007) US/O_3 combination degradation of aniline in aqueous solution. Ultrasonics Sonochemistry 14: 84-88.

[16] Sun Y, Pignatello J.J (1993) Photochemical reactions involved in the total mineralization of 2,4-D by $Fe^{3+}/H_2O_2/UV$. Environmental Science & Technology. 27: 304-310.

[17] Sun J.H, Sun S.P, Fan M.H, Guo H.Q, Qiao L.P, Sun R.Z. (2007) A kinetic study on the degradation of p-nitroaniline by Fenton oxidation process. J. Hazardous Materials. 148: 172-177.

[18] Walling C (1975) Fenton's reagent revised. Accounts of Chemical Research 8: 125-131

[19] Zhang H, Fei C, Zhang D, Tang F (2007) Degradation of 4-nitrophenol in aqueous medium by electro-Fenton method. J. Hazardous Materials 145: 227-232.

10

Setting Up of Risk Based Remediation Goal for Remediation of Persistent Organic Pesticides (Pesticide-POPs) Contaminated Sites

Tapan Chakrabarti
National Environmental Engineering Research Institute
India

1. Introduction

1.1 Persistent Organic Pollutants (POPs)

Persistent Organic Pollutants (POPs) is a common name of a group of pollutants that are semi-volatile, bioaccumulative, persistent and toxic. POPs comprising of pesticides, industrial chemicals and unintentionally produced POPs are toxic chemicals that adversely affect human health and the environment around the world (Vallack et al.,1998; Mocarelli &Tallman,1998 & Jones &de Voogt,1999). Since these pollutants can be transported by wind (natural ,human barriers and perturbation , grass in the desert ,cities and thermal updrafts, etc), water (ssurface, rain, ground, pumped and/or anthropogenically modified chemicals) and human intervention (waste-water conduits, drainage ditches, roadways, railways, irrigation, ponding, physical transformation physical collection and concentration, human species transformation) (Sw-846,USEPA,2007) , most POPs generated in one country can affect people and wildlife far from where they are used and released. These chemicals persist for long periods of time in the environment and can accumulate and pass from one species to the next through the food chain.

POPs can be deposited in marine and freshwater ecosystems through effluent releases, atmospheric deposition, runoff, and other means. As POPs have low water solubility, they bond strongly to particulate matter in aquatic sediments. As a result, sediments can serve as reservoirs or "sinks" for POPs. When sequestered in these sediments, POPs can be taken out of circulation for long periods of time. If disturbed, however, they can be reintroduced into the ecosystem and food chain, thereby potentially becoming a source of local, and even global contamination.

To address this global concern, there exists a groundbreaking United Nations' treaty in Stockholm, Sweden (May 2001). Under the treaty, known as the Stockholm Convention, countries agree to reduce or eliminate the production, use, and/or release of 12 key POPs (UNEP Tookits on POPs,1999,2001,2003,2005) which are shown in **Table 1**.

The Convention specifies a scientific review process that could lead to the addition of other POPs chemicals of global concern. The list of nine new POPs added to the Stockholm Convention in May, 2009 is shown in **Table 2**.

Sr. No.	POPs Identified
POPs-Pesticides	
1	Aldrin [1]
2	Chlordane [1]
3	Dichlorodiphenyl trichloroethane (DDT)[1]
4	Dieldrin[1]
5	Endrin[1]
6	Heptachlor[1]
7	Hexachlorobenzene [1,2]
8	Mirex[1]
9	Toxaphene[1]
POPs-nonpesticides	
10	Polychlorinated biphenyls (PCBs) [1,2]
11	Polychlorinated dibenzo-p-dioxins[2](dioxins)
12	Polychlorinated dibenzofurans[2] (furans)

1.   Intentionally produced.
2.   Unintentionally Produced - Result from some industrial processes and combustion.

Table 1. List of 12 POPs identified by the Stockholm Convention of May 2001

POPs	Usage
Alpha hexachlorocyclohexane	Pesticide, produced as byproduct of lindane
Beta hexachlorocyclohexane	Pesticide, produced as byproduct of lindane
Chlordecone	Pesticide, agricultural use
Hexabromobiphenyl ether	Flame retardant
Hexabromodiphenyl ether and heptabromodiphenyl ether	Flame retardant, recycling of articles containing these chemicals is allowed
Lindane (Gamma hexachlorocyclohexane)	Pesticide, for control of head lice and scabies as second line treatment
Pentachlorobenzene	Pesticide, unintentionally produced POPs
Perfluorooctane sulfonic acid, its salts and perfluorooctane sulfonyl fluoride	Industrial chemical: Photo-imaging, photo-resist and anti-reflective coatings for semi-conductor and liquid crystal display (LCD) industries, etching agent for compound semi-conductors and ceramic filters, aviation hydraulic fluids, metal plating (hard metal plating and decorative plating), certain medical devices (such as ethylene tetrafluoroethylene copolymer (ETFE) layers and radio-opaque ETFE production, in-vitro diagnostic medical devices, and CCD colour filters), fire-fighting foam, insecticides for control of fire ants and termites, electric and electronic parts for some colour printers and colour copy machines, chemically driven oil production, carpets, leather and apparel, textiles and upholstery, paper and packaging, coatings and coating additives, rubber and plastics
Tetrabromodiphenyl ether and pentabromodiphenyl ether	Flame retardant, recycling of articles containing these chemicals is allowed

Table 2. Nine new POPs added to the Stockholm Convention in May, 2009

An attempt is now being made to include endosulfun in the list of POPs.

1.2 Remediation requirement for obsolete pesticides stockpiles and contaminated sites

Obsolete pesticide stocks refer to pesticides that have been banned or whose shelf life has expired. Many international organizations are working on the issue of obsolete pesticide stocks. These include FAO, UNEP Chemicals and the Secretariat of the Basel Convention, UNIDO (United Nations Industrial Development Organization), industry associations and NGOs (non-governmental organisations) dealing with environment. Approximately 20,000 tons of obsolete pesticides are located in Africa and in the Middle East, often in containers that leak toxic waste into the environment (Fitz,2000). While exact quantities are unknown, large stockpiles also exist in Central Eastern European Countries (CEEC) and the New Independent States (NIS) (UNEP GEF Draft Report, 2002). The risks associated with large-scale storage of compounds pose a particular environmental risk. The principal uncertainly in terms of these obsolete stocks is characterization in terms of POP content. Little is known of the composition of the waste materials and it must be recognized that, within the 'cocktail' of possible chemicals, a variety of substances will be present in unknown amounts. These could represent locally and regionally important on-going primary source inputs of compounds to the environment.

A contaminated site can be defined as an area of the land in which the soil or any groundwater lying beneath it, or the water or the underlying sediment, contains a hazardous waste, or another prescribed substance in quantities or concentrations exceeding prescribed risk based or numerical criteria or standards or conditions.

Article 6 of the Stockholm Convention describes measures to reduce or eliminate releases from stockpiles and contaminated sites.

This chapter will focus on the risk assessment prior to remediation of sites contaminated with persistant organic persticides [DDT, aldrin, chlordane, dieldrin, endrin, heptachlor, hexachlorobenzene (HCB), mirex, toxaphene, and hexachlorohexane (HCH)]. The pesticide contaminated sites include pesticide manufacturing sites, pesticide formulation sites and other sites, including storage facilities and aerial application facilities (Fitz,2000;NNEMS,2000). Due to the complexity of these sites, it is difficult to make generalizations regarding the risk assessment for the types of contamination present and the remediation activities that were/ are being chosen.

The readers of this chapter are suggested to go through the UNIDO document titled "Persistent Organic Pollutants: Contaminated Site Investigation and Management Toolkit (2010) available on UNIDO site for free download (Contaminated Site Ttoolkit, 2010). This Toolkit aims to aid developing countries with the identification, classification and prioritization of POPs-contaminated sites, and with the development of suitable technologies for land remediation in accordance with best available techniques/best environmental practices (BAT/BEP). The Toolkit focuses exclusively on the 12 POPs listed in **Table 1**. The nine POPs recently added to the Stockholm Convention (listed in **Table 2**) are not included because there are still significant scientific challenges and unknowns associated with them. The Toolkit could be used both as a training tool and as a self-directed manual and resource document for decision-makers, practitioners and a range of other stakeholders. Pertaining to POPs contaminated site management

The readers may also peruse the Alberta Tier 1 Soil and Groundwater Remediation Guidelines (Alberta Tier 1, 2010) encompassing generic remediation guidelines for achieving equivalent land capability. Site-specific guidelines for achieving equivalent land capability can be developed using a Tier 2 approach (Alberta Tier 2,2010).

2. Site prioritization for risk assessment

2.1 Contaminated site prioritization

The purpose of the site prioritization is to classify contaminated sites based on risk assessment. In this section, two semi-quantitative tools are presented that allow the user to determine which sites should be assessed and then to prioritize sites based on their potential for causing unacceptable risks to humans and/or to natural environment. These are pre-screening tool and prioritization tools.

The pre-screening tool determines, through inventorization approach, as to whether the site has a history of activity leading to pesticide-POPs contamination or whether there are other reasons to believe that contaminants have been present at the site

The tool aims to gather contaminant characteristics, off-site migration potential, exposure and socio-economic factors. Then, based on the information obtained , the sites should be classified into the following categories:

- Class 1 – High priority for risk assessment
- Class 2 – Medium-high priority for risk assessment
- Class 3 – Medium priority for risk assessment
- Class 4 – Low priority for risk assessment
- Class N – Not a priority for risk assessment

The site prioritization tools are based on principles derived from the Canadian National Classification Tool for contaminated sites (CCME 2008).

The following web resources are helpful for designing a sampling program: http://www.ccme.ca/assets/pdf/pn_1101_e.pdf (PDF file) .

While the tools are applicable to any contaminated site, a greater emphasis has to be put on pesticide- POP's related contaminant issues.

2.2 Setting up of risk-based standards by regulatory agency
2.2.1 Risk based approach

The assessment before procedures and of contaminated sites is to follow the human health and environmental risk based approaches delineated in the Procedures for the Use of Risk Assessment under Part XV.1 of the Environmental Protection Act ,Ontario Ministry of the Environment, Canada. The Contaminated Sites Monograph Series, The Health Risk Assessment and Management of Contaminated Sites, and Australian Standard AS4482 – Guide to the sampling and investigation of potentially contaminated soil and the equivalent National Environment Protection Measure (NEPM) Guidelines(a,b,c,d,e,f and g,1999) are some of the important references for consultants carrying out contaminated site investigations.

Land becomes contaminated when there is spillage, leakage or disposal of pesticide POPs to the ground. Soil at or below the ground surface, and sometimes groundwater, as well, may be contaminated depending on the subsurface conditions. To determine objectively if a piece of land is contaminated, certain standards would need to be put in place. Generally developing countries have no locally-derived standards for land contamination assessment. These countries adopt standards from developed country(ies) which has/have very different conditions. Therefore there is a need to develop contaminated land standards that are tailor-made for local conditions.

The United States (US) Environmental Protection Agency (USEPA) pioneered the application of chemical risk assessment principles and procedures to evaluate contaminated sites under their Superfund Program in the 1980s. Other countries (mainly Canada, Australia and some European countries including the Netherlands) followed the US footsteps and began developing their own risk-based standards in the 1990s by making reference to the US approach. http://www.epd.gov.hk/epd/english/boards/ advisory_council/ files/ACE_Paper_18-2006.pdf)

The risk-based approach means that contaminated land will be managed by considering the nature and extent of the potential risk it poses as a result of the receptors' exposure to chemicals in the soil and/or groundwater. This basically acknowledges that there is an acceptably low level of exposure to contaminants, which poses negligible risk. Choosing the level of negligible risk is a very important decision in the derivation of risk-based standards. The risk levels usually considered for protection of public health are:

- An excess lifetime cancer risk of 1 in 10^6 for carcinogens
- Actual intake must be less than the safe dose for non-carcinogens.

These risk limits are in line with the international practice and are at the conservative end of the range of risk limits adopted worldwide. For example, it is noted that the risk limit of 1 in 10^6 has also been adopted by some countries such as the US, the Netherlands and Canada, etc. while the UK has used a higher risk of 1 in 10^5.

2.2.2 Health risk assessment

Establishment of Base Line Human Health Risk Assessment (BHRA)

The risk assessment procedures, developed and documented by US EPA, are still subject to improving, especially concerning human health (US EPA 1989; 1991a; 1995; 1996a; 1996b; 2001a; 2001b; 2002).

Generally, the purpose of base line human health risk assessment (BHRA) is to:

- Assess potential risks to human health
- Determine the need for remedial action
- Determine measures needed to eliminate or mitigate health and environmental effects.

BHRA is an analysis of the potential adverse health effects caused by exposure to hazardous substances released from a site in the absence of actions to control or mitigate these releases (i.e., under an assumption of no action) (US EPA 1989).

To estimate BHRA, the following steps are undertaken:

- Data collection and evaluation
- Selection of indicator chemicals
- Exposure assessment
- Toxicity assessment
- Risk characterization.

Data collection and evaluation

An objective of the data collection and evaluation step is to produce data that can be used to assess risks to human health. This step includes:

- Review of available site information
- Consideration of modeling parameter needs
- Collection of background data (samples not influenced by site contamination)
- Preliminary identification of potential human exposure

- Development of an overall strategy for sample collection
- Identification of analytical needs
- Collection and evaluation of data
- Development of a data set that is of acceptable quality for the risk assessment.

Human health implications of POPs

The implications of chronic and acute exposures to POPs are not fully understood. Laboratory investigations and environmental impact studies in the natural environment have indicated that POPs exposure can result in endocrine disruption, reproductive and immune dysfunction, brain and nervous system disorders, developmental disorders and cancer. Some organochlorine chemicals are likely carcinogenic by promoting the formation of tumors. Six of the 9 pesticide-POPs, identified in the *Stockholm Convention* , are classified as *possibly* carcinogenic to humans. The remaining three - endrin, dieldrin and aldrin are classified by WHO as highly hazardous (class 1b) on the basis of their acute toxicity to experimental animals.

Fetuses and infants are particularly vulnerable to pesticide POPs exposure due to the transfer of these POPs from the mother during critical stages of development. Exposure during development has been linked to reduced immunity (and increased infections), developmental abnormalities, brain and nervous system impairment, and cancer and tumor induction or promotion in infants and children. There may also be a link to human breast cancer.

Cancer risk

The International Agency for Research on Cancer identifies most of the 12 POPs targeted by the Stockholm Convention as presenting a potential carcinogenic risk to humans, as described in the **Table 3** below.

Sr. No.	IARC Classification	POPs
1	**Group 1:** The agent (mixture) is carcinogenic to humans	• 2,3,7,8-Tetrachlorodibenzo-para-dioxin (TCDD)
2	**Group 2A:** The agent (mixture) is probably carcinogenic to humans	• Mixtures of polychlorinated biphenyls (PCB)
3	**Group 2B:** The agent (mixture) is possibly carcinogenic to humans	• Chlordane • DDT • Heptachlor • Hexachlorobenzene • Mirex • Toxaphene (mixtures of Polychlorinated camphenes)
4	**Group 3:** The agent (mixture or exposure circumstance) is unclassifiable as to carcinogenicity in humans	• Aldrin, Dieldrin and Endrin • Polychlorinated dibenzo-para-dioxins (other than TCDD) • Polychlorinated dibenzofuran

Source: http://www.chem.unep.ch/gpa_trial/02healt.htm

Table 3. POPs posing potential carcinogenic risk to humans

Possible human exposure pathways

Humans can be exposed to pesticide-POPs through diet, occupation, accidents and both the indoor and outdoor environments. Exposure to these POPs can either be a short-term exposure to high concentrations (acute) or long-term exposure to lower concentrations (chronic).

Acute exposure to pesticide-POPs can occur during production and application and industrial accidents. In addition, exposure to chlorinated pesticides can occur both from accidental ingestion of treated seeds or via poor handling or application processes. Presently, pesticide poisoning is mainly attributable to aldrin, dieldrin, HCB and chlordane.

Chronic exposure occurs most commonly via dietary exposure pathways. Due to their tendency to bio-accumulate, longer-term human exposure to the pesticide- POPs identified in the Stockholm Convention is generally via food. Foods containing the greatest concentrations of POPs include the fatty tissues of animals and edible oils. The contamination of food, including breast milk, by POPs is of worldwide concern (Stober,1998).

Toxicity assessment

Toxicity assessment is based on available scientific data on potential adverse health effects of the contaminants in humans, which are usually compiled in the form of a toxicological profile for each contaminant. This step also includes also identification of important measures of toxicity, i.e., reference doses (RfDs) to evaluate non- carcinogenic effects, and cancer slope factors (CSFs) for carcinogenic effects.

RfDs and CSFs have been developed by the US EPA and published in the Integrated Risk Information System (IRIS) (IRIS 2003), and Health Effects Assessment Summary Tables (HEAST) databases. IRIS is recommended as a preferred source of toxicity information. HEAST is used when data are not available in IRIS (US EPA 1989; 2003).

The US EPA has also developed provisional values of RfDs and CSFs, which are used for specific purposes (US EPA 2003). If no RfDs and CSFs are available, the chemicals can be evaluated only qualitatively.

Exposure assessment

The exposure assessment stage estimates the magnitude of actual and/or potential human exposure, the frequency and duration of exposure, and pathways by which humans are potentially exposed (USEPA 1989).

The exposure assessment proceeds with the following steps:

Step 1. Characterization of exposure setting - the physical environment of the site and the potentially exposed populations are characterised.

Step 2. Identification of exposure pathways - chemical sources and mechanism of chemical release, transport media (e.g., soil, air, groundwater), exposure points as well as exposure routes (e.g., ingestion, inhalation, dermal contact) are identified in this step; an exposure pathway describes the course a chemical or physical agent takes from the source to the exposed individual (e.g., ingestion of contaminated schoolyard soil by children).

Step 3. Quantification of exposure - exposure concentrations of contaminants are estimated and pathway-specific intakes are calculated.

During this step, site-specific exposure scenarios are developed for both current and/or intended future land use patterns (e.g., residential, commercial/industrial, recreational).

Results of the exposure assessment are pathway-specific contaminant intakes, under developed exposure scenarios. Standard intake equations and suggested values of exposure parameters are provided by the US EPA; however, site-specific factors and expert judgment can influence the final selection thereof.

In the classical approach, exposure parameters, such as body weight, exposure duration, ingestion or inhalation rates, can be selected to estimate "reasonable maximum exposure" (RME), defined as the highest exposure that is reasonably expected to occur at a given site (US EPA 1989). The goal of RME is to combine upper-bound and mid-range exposure factors in the equation so that the result represents an exposure scenario that is both protective and reasonable, not the worst possible case (US EPA 1991b).

The quantification of exposure is based on an estimate of the average daily intake, i.e., the average amount of the contaminant entering the receptor's body per day.

The considered human receptors are strictly related to defined land use patterns, e.g., adult receptors under the industrial land use, and children and adults under residential/recreational land uses.

The generic equation for calculating chemical intakes is as follows:

$$DI = C \times (IR / BW) \times (EF \times ED / AT) \tag{1}$$

where:

DI =   daily intake of chemical (mg/kg-d)
C =   concentration of chemical in an environmental medium (e.g., mg/kg for   soil or food, mg/L for water, mg/m^3 for air)
IR =   intake rate of the environmental medium (e.g., kg/day for food or soil, L/day for water, m^3/day for air)
BW =   body weight (kg)
EF =   exposure frequency (days/yr)
ED =   exposure duration (years)
AT =   averaging time (days)

It may be noted that the term IR/BW is a description of the basic contact rate with a medium (e.g., L of water per kg body weight per day) and the second term (EF×ED/AT) adjusts for cases where exposure is not continuous. For example, if a person was exposed for 50 days/year for 20 years of a lifetime (70 years), the value of this term would be $50/365 \times 20/70 = 0.039$.

There is often wide variability in the amount of contact between different individuals within a population. Thus, human contact with an environmental medium is best thought of as a distribution of possible values rather than a specific value. Usually, emphasis is placed on two different points of this distribution:

Average or Central Tendency Exposure (CTE)

CTE refers to individuals who have average or typical intake of environmental media.

Upper bound or Reasonable Maximum Exposure (RME)

RME refers to people who are at the high end of the exposure distribution (approximately the 95th percentile). The RME scenario is intended to assess exposures that are higher than average, but are still within a realistic range of exposure.

As the calculations of CTE and RME risk are done using single numbers (point estimates) for each input value, this approach is usually referred to as the point estimate method. In some cases, the risk assessor may wish to describe each exposure parameter not by a single number but as a distribution. This is referred to as probabilistic risk assessment (PRA). In this case, computations require computer-based methods (Monte Carlo simulation) and the output is also a distribution rather than a point estimate. This approach provides a more complete description of the range of exposures that occur in the exposed population and also helps increase the accuracy of combining exposure levels across different pathways.

In some cases, human exposure may be measured directly (biomonitoring) rather than calculated based on assumed exposure parameters. For example, exposure to lead is often evaluated by measuring the amount of lead in blood, and exposure to arsenic is often evaluated by measuring the amount of arsenic in urine or in hair. While direct measurement bypasses many of the uncertainties associated with calculating human exposure, this approach is limited by providing data only on current conditions. In addition, if exposure is occurring from more than one source, direct measurement does not distinguish between the sources.

Equations for exposure pathway to contaminated soil for outdoor and indoor workers are in the Tables published by USEPA (USEPA, BJC/OR-271, 2006.) URL: http://rais.ornl.gov/ homepage.

Risk characterization

Risk characterization combines toxicity assessment with exposure assessment, in order to quantify risks posed by a contaminated site under a given set of conditions.

Risk characterization is considered separately for carcinogenic and non-carcinogenic effects, and includes identification of sources of uncertainty. Chemicals, which produce both non-carcinogenic and carcinogenic effects are evaluated in both groups.

Risks are quantified under the present site conditions for present and/or future exposure scenarios relevant to the land use pattern. Risk characterization should also include a discussion on accompanying uncertainties.

Non-cancer risk

Potential non-cancer risks are evaluated by comparison of the estimated contaminant intakes from each exposure route (oral, dermal, inhalation) with the relevant RfD to produce the hazard quotient (HQ), defined as follows (US EPA 1989):

$$HQ\text{-ingestion and dermal} = CDI/RfD$$
$$HQ\text{-inhalation} = CDI/RfC$$

where:

HQ:Hazard Quotient (unitless),

CDI: Chronicl Daily Intake (mg/kg/day),

RfD: Reference Dose (mg/kg/day).

RfC : Reference Concentration

The hazard quotient assumes that there is a level of exposure (i.e., RfD/RfC) below which it is unlikely to experience adverse health effects, even for sensitive populations. If the HQ exceeds unity (a value of 1), there may be a concern for potential non- carcinogenic effects.

To assess the overall potential for non-carcinogenic health effects posed by more than one chemical, the HQs calculated for each chemical are summed (assuming additivity of effects), and expressed as a Hazard Index (HI) (US EPA 1989).

$$HI = HQ1 + HQ2 + ... + HQn \tag{2}$$

In cases where the non-cancer HI does not exceed unity (HI<1), it is assumed that no chronic risks are likely to occur at the site (US EPA 1989). If the HI is higher than unity, as a consequence of summing several hazard quotients, the compounds are segregated by effects, target organs, and by mechanism of action and separate HIs are derived for each group.

Because of the potential for different health effects/target organs via oral/dermal and inhalation exposures, these exposures are evaluated separately (US EPA 2002). To assess the overall potential for non-carcinogenic effects, posed by several exposure pathways, HIs for each exposure pathway contributing to exposure of the same individual or subpopulation are summed up and expressed as a total hazard index (HI Tot). When HI Tot exceeds unity, there may be concern for potential non-cancer health effects.

Quantitative risk assessment

Under the residential and recreational scenarios, i.e., scenarios which refer to different group receptors (children, adults), HIs are generated separately for children and adults.

Cancer risk

Cancer risks are estimated as the incremental probability of an individual developing cancer over a lifetime as a result of exposure to the potential carcinogen (i.e., incremental or excess individual lifetime cancer risk). The following linear low-dose carcinogenic risk equation is used for each exposure route (US EPA 1989):

$$\text{Excess Lifetime Cancer Risk (ELCR)-ingestion \& dermal} = \text{CDI} \times \text{slope factor (CSF)} \quad (3)$$

$$\text{Excess Lifetime Cancer Risk (ELCR) - inhalation} = \text{CDI} \times \text{unit risk factor (URF)} \quad (4)$$

$$\text{Cancer Risk} = \text{CDI} \times \text{CSF} \quad (5)$$

where:

CDI: Chronic Daily Intake averaged over 70 years (mg/kg/day),

CSF/URF:: Cancer Slope/Unit Risk Factor (mg/kg/day) 1 ; a plausible upper-bound estimate of the probability of a response per unit intake of a chemical over a lifetime.

CDI and CSF/URF represent the same exposure route (i.e. oral, dermal and inhalation CDIs are multiplied by oral, dermal and inhalation CSFs/URF, respectively). The risk number represents the probability of occurrence of additional cancer cases. For example, if it is expressed as 1E-06, it means that one additional case of cancer is expected in a population of one million people exposed to a certain level of a given chemical over their lifetime.

If a site has multiple carcinogenic contaminants, cancer risks for each carcinogen are added (assuming additivity of effects), and the cancer risk for each exposure pathway is calculated. For multiple exposure pathways, the total cancer risk is calculated by summing up the pathway-specific cancer risks:

Risks in the range of 1E-06 to 1E-04 are generally accepted by regulatory agencies, e.g., US EPA (US EPA 1990; 1991a; 1991c). A risk-based remedial decision can be superseded by the presence of a non-carcinogenic impact or environmental impact requiring action at the site. Remedial action is generally required at a site, when a cumulative carcinogenic risk exceeds 100 in a million (1E-04, excess cancer risk) or the cumulative non-carcinogenic HI exceeds 1, based on RME assumptions (US EPA 1991a; 1991c). If the cumulative risk is less than 1E-04, action generally is not required, but may be warranted if a risk-based chemical-specific standard (e.g., drinking water standards) is violated. Setting up 1E-06 risk level for individual chemicals and pathways should generally lead to cumulative site risks within the range of 1E-06 to 1E-04 for the combinations of chemicals.

Under the scenarios, which refer to both receptors – a child and an adult (i.e., residential and recreational), cancer risks are calculated separately for these receptors, and then summed up to yield the total cancer risk for the aggregate resident/recreational user.

2.3 Dealing with biased data

The basic unit of a risk assessment is an exposure unit, and the key description of exposure is the arithmetic mean concentration within an exposure unit. If the data collected from within an exposure unit are either random or systematic, the methods for computing the mean (and confidence limits around the mean) are relatively straightforward. However, in some cases, the data available are not random or systematic, but are biased. That is, more samples are collected from areas with high concentrations than with low concentrations. This unequal sampling density poses a difficulty in computing the mean, but techniques are available for adjusting for this issue. Important guidance documents on how to make these adjustments include the following:

Spatial Analysis and Decision Assistance (SADA) Software Home Page GeoSEM Software (Syracuse Research Corporation)

2.4 Probabilistic Risk Assessment (PRA)

Equations for computing human exposure contain a number of terms that are inherently variable. For example, not all people have the same body weight. Rather, there is a distribution of body weights across different people. The same is true for intake rates, exposure frequencies, and exposure durations. If data are available to describe the distribution of each of these terms, then a mathematical method is needed to combine the distributions.

While there are a number of different methods available, the most common and convenient is Monte Carlo simulation. In this approach, each term in the exposure model is described by a distribution rather than a single value. The computer draws a value at random from each distribution, computes the exposure, and saves the value. This process is repeated many times, resulting in a distribution of exposure values. This distribution provides a more complete description of exposure than the point estimate approach and helps ensure that values selected for CTE and RME exposures are realistic. Key guidance documents dealing with PRA include the following:

- RAGS III Part A: Process for Conducting Probabilistic Risk Assessment (OSWER 9285.7-45, December 2001)
- Note: In particular, see Chapter 3 - Using Probabilistic Analysis in Human Health Assessment (PDF) (27 pp, 2MB).
- Guiding Principles for Monte Carlo Analysis (EPA/630/R-97/001, March 1997)
- Policy for Use of Probabilistic Analysis in Risk Assessment at the U.S. Environmental Protection Agency (May 1997)

2.5 Biomonitoring

In some cases, biomonitoring may be a useful tool to help evaluate current exposure levels at a site. This requires that a population of humans are present at the site and that there is a method available for measuring the level of exposure in the population. In general, the results of the biomionitoring may be compared to other (reference) populations to help understand the magnitude of the site-related exposure, and/or may be compared to health-based guidelines for the maximum level of exposure that is considered acceptable. Important guidance documents on planning, performing, and interpreting biomonitoring studies are presented below.

- Criteria for Evaluating Blood Lead (PDF) (Region 8 Guidance RA-07, September 1995) (22 pp, 1.6MB)
- Sample Analysis and Quality Assurance Plan for Urinary Arsenic and Blood Lead Among Residents of VBI70 Neighborhoods (PDF) (Region 8, June 2002) (27 pp, 333K)
- Experience Using Filter Paper Techniques for Whole Blood Lead Screening in a Large Pediatric Population (PDF) (8 pp, 187K) (J.A. Collins and S.E. Puskas, MEDTOX Laboratories, Inc., Saint Paul, MN)

3. Development of site-specific Health-based Remedial Goals (HBRGs)

3.1 HBRGs

Health-based remedial goals (HBRGs), termed also risk-based concentrations [RBCs,(http://www.image-train.net/products/papers/ASC3_EW_RBA.pdf)], are concentration levels for individual chemicals that correspond to target risk (TR), i.e., a specific cancer risk level (e.g., $1E^{-06}$) or hazard quotient (HQ) or hazard index (HI) (e.g., less than or equal to 1) (US EPA 1991a). RBCs are usually calculated under all developed scenarios for the purpose of guiding remedial activities at a site; they are used during analysis and selection of remedial alternatives.

There are two methods for calculating RBCs. The first method (Method 1) is a simplified method based on site- specific exposure data (US EPA 1995). This method uses the ratio between the target risk and calculated risk due to a specific chemical in a given medium:

$$\frac{C}{\text{Calculated Risk}} = \frac{RBC}{\text{Target Risk}} \qquad (6)$$

where:
C: Chemical Concentration in soil or groundwater
RBC: Risk-Based Concentration (oral/dermal or inhalation).
Rearranging this equation, RBC is calculated as follows:

$$RBC = C \frac{\text{Target Risk}}{\text{Calculated Risk}} \qquad (7)$$

RBCs are calculated for both carcinogenic and non-carcinogenic substances, and only for those contaminants for which the calculated site-specific risk is above acceptable risk levels (target risk). For carcinogens, RBCs can be calculated for target cancer risks of 1E-06, 1E-05 or 1E-04. Concerning non-carcinogenic risk,target HQs of 0.1 or 1 can be substituted for target risk, and the calculated HQs substituted for calculated risks.

Under industrial scenario, RBCs are estimated for adult receptors, and under the residential/recreational scenarios - separately for child and adult receptors for non-carcinogenic effects, and for an aggregate resident/recreational user for carcinogenic

Setting Up of Risk Based Remediation Goal for Remediation of Persistent Organic Pesticides (Pesticide-POPs) Contaminated Sites

193

effects. According to the US EPA recommendations, RBCs are calculated separately for oral/dermal and inhalation exposures, because of the potential for different health effects (target organs) via these routes (US EPA 2002). If both carcinogenic and non-carcinogenic RBCs are calculated for a given contaminant, and for both oral/dermal and inhalation exposures, then lowest of these values should be applied as the preliminary remedial goal.

Concerning non-carcinogens, if more than one chemical affects the same target organ/system, RBCs calculated for those chemicals should be divided by the number of chemicals present in the group. In that way, RBCs are adjusted to reflect the potential for additive risks:

$$ARBC = RBC/n \tag{8}$$

where:

ARBC: Risk-Based Concentration adjusted for exposure to multiple contaminants with the same target organs/effects

RBC: Risk-Based Concentration for an individual non-carcinogen

n: Number of contaminants with the same target organs/effects.

RBCs can also be calculated by rearrangement of standard risk equations, separately for combined oral and dermal exposures, and for inhalation exposure by a receptor within the used scenario (US EPA 2002).

In summary, application of risk-based approach to contaminated land assessment and remediation allows to:

- Determine the needs for remedial action, aimed at reducing risk,
- Determine preliminary remedial goals based on the protection of human health,
- Provide a basis for the selection of an appropriate remedial option
- Facilitate making decision on appropriate corrective actions at the site.

3.2 Health concerns

Persistent pesticides pose a threat to the well-being of the environment and to human health. The solid organochloride insecticides are known to accumulate in human adipose tissue. Some of these insecticides, including chlordane, can even be absorbed dermally. Other health problems caused by exposure to the solid organochloride insecticides are convulsions, a hyperexcitable state of the brain and a predisposition to cardiac arrhythmia. Eating wheat treated with hexachlorobenzene, another organochloride insecticide, has been associated with human dermal toxicity, which can result in blistering of the skin. Although not all organochlorine insecticides are considered POPs, many of them are among the compounds on the UNEP's list of persistent organic pollutants, including aldrin, chlordane, DDT, dieldrin, endrin, heptachlor, hexachlorobenzenes, mirex and toxaphene.

The assessment of health effects of contamination is to be made with reference to the human health-based investigation levels for various settings described in Contaminated Sites Monograph Series No 5, 1996 as incorporated in Appendix 9. The assessment of environmental impacts is to include reference to the ANZECC/NHMRC Environmental Investigation Thresholds in Appendix 9.1.

In urban residential settings where sensitive ecological receptors are not present, assessment should address:

- Potential health risks to occupants
- The capacity of the soil to support a normal ornamental domestic garden without significant phytotoxic effect.

In this process, reference may be made to the contemporary background metal levels for Queensland horticultural soils determined in studies by the Department of Natural Resources (full reference in Appendix 9.2). The chemical form of the contaminant and its mobility characteristics will be essential components of assessments.

Complex health risk assessments are to be undertaken by qualified professionals using nationally accepted health risk assessment methodology when a significant exposure risk exists. A suitable module on Site-Specific Health Risk Assessment and Management of Contaminated Sites is to be refererred to, if available.

Environmental risk assessment is to be conducted on a site-specific basis when contamination levels exceed background. The characteristics of the contaminant (including chemical form, mobility, leachability and bioavailability) and the exposure routes to local receptors should be identified.

4. Integrating risk assessment with contaminated site management

Risk is governed by the contaminants present on the site, pathways through which these contaminants reach the receptors and the receptors who are usually the site users. A conceptual site model, encompassing all the three site attributes , usually helps to focus on risk. Construction of a contaminant-pathway-receptor model is the first step of risk assessment.

Since remediation through risk management deals with eliminating or controlling one or more of the three risk components: (i) contaminant, ii) exposure pathway, and iii) receptor, remediation becomes the proactive risk management solution. The remediation measures include:

- Source control
- Site stabilization and decontamination to the extent required by the Regulatory Agency for a specific site-use purpose(s).
- Alternative forms of risk management on a contaminated site, such as exposure barriers, administrative controls and/or partial remediation, may be acceptable to a regulatory agency in certain cases.

Remediation , either on-site (in-situ) or off-site (ex-situ) can employ one method, or a combination of the available physical, chemical and biological methods.Sometimes, it is not possible to remove the contaminants or exposure routes due to technical or economic or environmental constraints, the last resort is to control the receptor's accessibility by relocations and imposing land use restrictions.

Long-term remediation strategies are intended to implement a comprehensive monitoring program that properly characterizes the baseline (pre-remediation) condition and monitors improvements to be achieved through targeted remediation. Long-term remedial measures focus on compliance with all regulatory standards applicable to all contaminated media (e.g., groundwater, soil, and soil vapour) present at the site.

The readers of this chapter are suggested to go through the UNIDO document titled "Persistent Organic Pollutants: Contaminated Site Investigation and Management Toolkit (2010) available on UNIDO site for free download. This Toolkit aims to aid

Setting Up of Risk Based Remediation Goal for Remediation of Persistent Organic Pesticides (Pesticide-POPs)
Contaminated Sites

195

developing countries with the identification, classification and prioritization of POP-contaminated sites, and with the development of suitable technologies for land remediation in accordance with best available techniques/best environmental practices (BAT/BEP).

5. References

Alberta Tier 1 Soil and Groundwater Remediation Guidelines Alberta Environment December 2010

Alberta Tier 2 Soil and Groundwater Remediation Guidelines Alberta Environment December 2010

Contaminated Site Toolkit (2010)
http://www.unidohttp://www.unido.org/fileadmin/user_media/Services/Environmental Management/Stockholm_Convention/POPs/toolkit/Contaminated site.pdf

Environmental Sampling (2007) SW-846 Update IV, DRAFT DOCUMENTATION

Fitz, N. (2000). Pesticides at Superfund Sites. Unpublished Data.

Jones, K.C. & de Voogt, P. (1999) Persistent Organic Pollutants (POPs): State of the Science. Environmental Pollution Vol.100 , pp. 209–221.

Mocarelli,P.&Taalman,R.D.F(1998) Controlling Persistent Organic Pollutants — What next?. Environmental Toxicology and Pharmacology Vol.63 , pp. 143–175.

National Environment Protection Measure (NEPM) Guidelines

a. Schedule B(4): Health Risk Assessment Methodology - Dec 1999

This Guideline incorporates aspects of the ANZECC Guidelines for the Laboratory Analysis of Contaminated Soil 1996, which were prepared in response to a recognised need for consistent procedures of soil analysis for environmental assessment of contaminated land.

The Guideline covers the philosophy behind the methods selected, it also comprises guidelines on the quality assurance procedures and techniques for sample preparation and describes methods for the analysis of physico-chemical properties, inorganics and organics in soil.

filename: ASC_NEPMsch__03_Lab_Analysis_199912.pdf

This document provides an approach to site-specific health risk assessment. Due to the complexity and scale of the health risk assessment process a concise 'cookbook' is not practicable. Similarly, the site-specific issues are often sufficiently complex and 'site-specific' for a particular site that a manageable and complete algorithm for decision-making cannot be drafted. The document provides a series of guidelines (and prescriptions) to assist the decision-making process. Where possible, the document is prescriptive about certain aspects of risk assessment.

filename: ASC_NEPMsch__04_Health_Risk_Assessment_199912.pdf

b. Schedule B(5): Ecological Risk Assessment - Dec 1999

The overall aim of this guideline is to promote a consistent, rational approach to ecological risk assessment of site contamination throughout Australia. Specifically, this document aims to provide a clear framework for ecological risk assessment for chemically contaminated soils that can be readily and consistently used by jurisdictional environmental agencies and risk assessors.

filename: ASC_NEPMsch__05_Ecological_Risk_Assessment_199912.pdf

c. Schedule B(6): Risk Based Assessment of Groundwater Contamination- Dec 1999

The purpose of this draft Guideline is to provide a framework for the risk based assessment of groundwater that may have been affected by site contamination. The general processes outlined for the assessment of contaminated groundwater are compatible with the Policy Framework and the site assessment processes shown in Schedule A of the NEPM. The aim of this process is to minimise the risk of adverse human health and environmental impacts arising from contaminated groundwater.
filename: ASC_NEPMsch__06_Groundwater_199912.pdf
d. Schedule B(7a): Health-Based Investigation Levels - Dec 1999
e. This guideline was published jointly with the National Environmental Health Forum (NEHF) and is part of a series of NEHF monographs.
This guideline discusses the general principles for deriving guidance values for health-based investigation levels and also explores the process applied to develop health-based investigation levels for soils.
filename: ASC_NEPMsch__07a_Health_Based_Investigation_Levels_199912.pdf
vvvSchedule B (7b): Exposure Scenarios and Exposure Settings - Dec 1999
This guideline was published jointly with the National Environmental Health Forum (NEHF) and is part of a series of NEHF monographs.
This paper focuses on the component of exposure scenarios which may be seen as exposure settings (or standard land uses), with some reference to the characteristics of the populations potentially exposed in those settings. The intention is to define more clearly a standard range of exposure settings which regulators and risk assessors could use as baseline cases, to improve consistency of assessments, and provide a sound basis for land use/planning and remediation decisions based upon such risk assessments.
filename: ASC_NEPMsch__07b_Exposure_Scenarios_199912.pdf
f. Schedule B(8): Community Consultation and Risk Communication-Dec. 1999
This guideline provides a systematic approach to effective community consultation and risk communication in relation to the assessment of site contamination. It is not intended to be prescriptive but is intended to be used as a tool for effective consultation by consultants and regulators and should also provide a useful reference for all stakeholders including industry, government, landholders and the wider community. It should be noted that, in addition to this Guideline, each State or Territory has its own regulatory requirements regarding notification of pollution to the appropriate regulatory agency.
filename: ASC_NEPMsch__08_Community_Consultation_199912.pdf
g. Schedule B(9): Protection of Health & the Environment During the Assessment of Site Contamination Dec 1999
NNEMS (2000) The Bioremediation and Phytoremediation of Pesticide-contaminated Sites Prepared by Chris Frazar National Network of Environmental Studies (NNEMS) Fellow Compiled June - August 2000).
Stober, J. (1998). Health effects of POPs, *Proceedings of the Subregional Awareness Raising Workshop on Persistent Organic Pollutants (POPs)*, pp. 11-14. Kranjska Gora, Slovenia.
UNEP Toolkits on POPs (1999,2001,2003,2005
Vallack,H.W.; Bakker D. J.; Brandt,I.; Brorström-Lundén,E.; Brouwer,A.; Bull,K.R. (1998); Gough,C.; Guardans,R.; Holoubek,I.; Jansson,B.; Koch,R.; Kuylenstierna,J.; Lecloux,A.; Mackay,D.; McCutcheon,P.; Controlling persistent organic pollutants-What next? Environ.Toxicol.Pharm.6(3).143-175.

Permissions

The contributors of this book come from diverse backgrounds, making this book a truly international effort. This book will bring forth new frontiers with its revolutionizing research information and detailed analysis of the nascent developments around the world.

We would like to thank Margarita Stoytcheva, for lending her expertise to make the book truly unique. She has played a crucial role in the development of this book. Without her invaluable contribution this book wouldn't have been possible. She has made vital efforts to compile up to date information on the varied aspects of this subject to make this book a valuable addition to the collection of many professionals and students.

This book was conceptualized with the vision of imparting up-to-date information and advanced data in this field. To ensure the same, a matchless editorial board was set up. Every individual on the board went through rigorous rounds of assessment to prove their worth. After which they invested a large part of their time researching and compiling the most relevant data for our readers. Conferences and sessions were held from time to time between the editorial board and the contributing authors to present the data in the most comprehensible form. The editorial team has worked tirelessly to provide valuable and valid information to help people across the globe.

Every chapter published in this book has been scrutinized by our experts. Their significance has been extensively debated. The topics covered herein carry significant findings which will fuel the growth of the discipline. They may even be implemented as practical applications or may be referred to as a beginning point for another development. Chapters in this book were first published by InTech; hereby published with permission under the Creative Commons Attribution License or equivalent.

The editorial board has been involved in producing this book since its inception. They have spent rigorous hours researching and exploring the diverse topics which have resulted in the successful publishing of this book. They have passed on their knowledge of decades through this book. To expedite this challenging task, the publisher supported the team at every step. A small team of assistant editors was also appointed to further simplify the editing procedure and attain best results for the readers.

Our editorial team has been hand picked from every corner of the world. Their multi-ethnicity adds dynamic inputs to the discussions which result in innovative outcomes. These outcomes are then further discussed with the researchers and contributors who give their valuable feedback and opinion regarding the same. The feedback is then collaborated with the researches and they are edited in a comprehensive manner to aid the understanding of the subject.

Apart from the editorial board, the designing team has also invested a significant amount of their time in understanding the subject and creating the most relevant covers. They scrutinized every image to scout for the most suitable representation of the subject and create an appropriate cover for the book.

The publishing team has been involved in this book since its early stages. They were actively engaged in every process, be it collecting the data, connecting with the contributors or procuring relevant information. The team has been an ardent support to the editorial, designing and production team. Their endless efforts to recruit the best for this project, has resulted in the accomplishment of this book. They are a veteran in the field of academics and their pool of knowledge is as vast as their experience in printing. Their expertise and guidance has proved useful at every step. Their uncompromising quality standards have made this book an exceptional effort. Their encouragement from time to time has been an inspiration for everyone.

The publisher and the editorial board hope that this book will prove to be a valuable piece of knowledge for researchers, students, practitioners and scholars across the globe.

List of Contributors

Deepa T.V., G. Lakshmi and Lakshmi P.S.
Amrita School of Pharmacy, Amrita Viswa Vidyapeetham, India

Sreekanth S.K.
St.James Medical Acadamy,Calicut University, India

Trine Lund and Hafizur Rahman
Norwegian University of Life Sciences, Norway

André Luiz Meleiro Porto, Gliseida Zelayarán Melgar, Mariana Consiglio Kasemodel and Marcia Nitschke
Universidade de São Paulo, Instituto de Química de São Carlos, Brazil

Rita Földényi
Institute of Environmental Sciences, University of Pannonia, Veszprém, Hungary

Imre Czinkota and László Tolner
Department for Soil Science and Agricultural Chemistry, Szent István University, Gödöllő, Hungary

Paul A. Horne and Jessica Page
IPM Technologies Pty Ltd, Australia

Marie-Pierre Rivière, Michel Ponchet and Eric Galiana
INRA (Institut National de la Recherche Agronomique), Unité Mixte de Recherche 1301, Interactions Biotiques et Santé Végétale, F-06903 Sophia Antipolis, France
CNRS (Centre National de la Recherche Scientifique), Unité Mixte de Recherche 6243, Interactions Biotiques et Santé Végétale, F-06903 Sophia Antipolis, France
Université de Nice-Sophia Antipolis, Unité Mixte de Recherche Interactions Biotiques et, Santé Végétale, F-06903 Sophia Antipolis, France

Rosalina González Forero
La Salle University, Colombia

H.K. Manonmani
Fermentation Technology and Bioengineering, Central Food Technological Research Institute, (Council of Scientific and Industrial Research), India

Chavalit Ratanatamskul
Department of Environmental Engineering, Chulalongkorn University, Thailand

Nalinrut Masomboon
Department of Interdisciplinary programs in Environmental Science, Graduate School, Chulalongkorn University, Thailand

Ming-Chun Lu
Department of Environmental Resources Management, Chia-Nan University of Pharmacy and Science, Taiwan

Tapan Chakrabarti
National Environmental Engineering Research Institute, India